THE ESSENCE OF

DIGITAL DESIGN

THE ESSENCE OF ENGINEERING SERIES

Published titles

The Essence of Solid-State Electronics
The Essence of Electric Power Systems
The Essence of Measurement
The Essence of Engineering Thermodynamics
The Essence of Power Electronics
The Essence of Analog Electronics

Forthcoming titles

The Essence of Optoelectronics
The Essence of Microprocessor Engineering
The Essence of Communications Theory

THE ESSENCE OF

DIGITAL DESIGN

Barry Wilkinson
University of North Carolina at Charlotte

An imprint of Pearson Education
Harlow, England · London · New York · Reading, Massachusetts · San Francisco
Toronto · Don Mills, Ontario · Sydney · Tokyo · Singapore · Hong Kong · Seoul
Taipei · Cape Town · Madrid · Mexico City · Amsterdam · Munich · Paris · Milan

Pearson Education Limited
Edinburgh Gate
Harlow
Essex CM20 2JE
England

and Associated Companies throughout the world

Visit us on the World Wide Web at:
http://pearsoneduc.com

Printed and bound in Great Britain by
Bookcraft, Bath

Library of Congress Cataloging-in-Publication Data

Available from the publisher

British Library Cataloguing in Publication Data

A Catalogue record for this book is available from
The British Library

ISBN 0-13-570110-4 (pbk)

---

10 9 8 7 6 5
06 05 04 03 02

To my wife, Wendy
and my daughter, Johanna

Contents

Preface

The purpose of this book is to present all the key aspects of logic design as a first course for undergraduate Electrical Engineering and Computer Science students. The material has been developed specifically for one semester or course module.

Chapter 1 introduces the concept of a digital system and its application areas. The concept of a logic signal, which has one of two values, is introduced. All digital systems use logic signals that take one of two values, 0 and 1, both for control operations and for numbers (binary numbers). The binary number representation and binary arithmetic are outlined, in so far as it applies in the design of a digital system. Being an "essence" book, excessive details on binary numbers such as how to convert numbers to and from different number representations is given only a passing mention. References are given throughout to texts that extend the presentation in the book.

The underlying mathematical basis for a digital system is Boolean algebra, an algebra which, in this context, defines operations on two-valued variables. Boolean operations are implemented with logic gates, making the logic gate the fundamental component of a digital system. Logic gates are described in Chapter 2 with the relevant Boolean algebra underpinning. At the end of the chapter, quite simple electronic circuit details of some logic gates are outlined. This section will be of most interest to electrical engineers. Computer science students could omit the section without loss of continuity.

Logic gates are called combinational logic circuits because their outputs depend upon a combination of input values. Chapter 3 explores how to design more complex "combinational logic" digital systems using logic gates. One of the key aspects in the design of digital systems is the number and types of gates used. Often a minimum number of gates is desired. Techniques that lead to a reduced number of gates is described, notably the Karnaugh map minimization technique. Minimization of very complex logic circuits is often done by a computer program. Techniques more applicable to computer programs can be found in the list of further reading.

Most digital systems contain memory that can store information about past events. The response of such digital systems can then depend upon the past events. The basic logic components having memory are flip-flops which are described in Chapter 4. Flip-flops form the basis of more complex "sequential logic" circuits which can create sequences of output values. One very common sequential logic circuit is the counter whose outputs follow numeric sequences such as binary increasing sequences. The design of counters is also described in Chapter 4. Knowledge of the internal design of flip-flops can be found in the list of further reading, though this detail is often unnecessary since most designs use prefabricated flip-flops or standard designs.

Chapter 5 continues the design of sequential circuits. The method of producing a design starting with a state diagram is described through the use of examples. All the designs considered are so-called synchronous sequential logic circuits, those which are controlled by an external clock signal. Synchronous sequential logic circuits form the vast majority of designs because of the simplicity of timing operations with a clock signal. The design of asynchronous sequential logic circuits, which do not use a synchronizing clock signal, is described in the list of further reading.

Chapter 6 is devoted to one logic device family, the programmable logic device (PLD). PLDs are logic components that have alterable internal connections which enable various logic designs to be implemented totally within the device. PLDs are widely used to reduce the number of components and the cost. PLDs have to be "programmed" to obtain the desired internal connections. This programming is specified using a PLD programming language as described in Chapter 6. Sufficient details of a PLD language are given to enable designs to programmed. Complete details of PLD languages can be found in the further reading.

The final chapter, Chapter 7, introduces the area of logic circuit testing to detect faults that might have occurred during manufacture or subsequently. Testing is extremely important but has been omitted from many previous texts on logic design. Proper consideration should be given to manufacturing a working system as well as obtaining a logically correct design, and testing should be part of the design process. Chapter 7 is intended as a overview of the important techniques. More details of testing can be found in specialized texts listed in the further reading section.

As can be seen from the description of the chapters, there are several pointers to where the material can be developed. A solutions manual is also available to instructors.

I wish to record my continued appreciation to Christopher Glennie of Prentice Hall for his guidance throughout the preparation of this book, and his work in obtaining constructive reviews of the manuscript. I also wish to thank Jacqueline Harbor for handling the manuscript in its final stages and all of the Prentice Hall production staff for their professionalism. Finally, I thank the anonymous reviewers for their efforts and helpful comments. I would greatly appreciate any further suggestions or corrections to be send to me at abw@uncc.edu so that they can be incorporated into subsequent printings of this text.

Barry Wilkinson

CHAPTER 1

Digital systems and the representation of information

Aims and objectives

The purpose of this chapter is to establish the elements of a digital system and its application areas. The main part of this chapter concentrates upon how information is represented inside a digital system. The concept of a logic signal having two values is introduced and it is shown how such two-valued signals can be used to represent both control actions and numbers.

1.1 The realm of digital systems

A digital system is often designed to satisfy two closely interrelated tasks:

1. To control apparatus
2. To perform calculations.

A digital system might perform some calculation and, on the basis of the result, take certain control actions. Examples of such digital "control" systems would be a system to control (a) industrial equipment and (b) an automobile.

1.1.1 Control applications

Let us start with a simple industrial control application of controlling the amount of material held in a hopper, as shown in Figure 1.1. The task here is to maintain material in the hopper between level L1 and level L2. The material is taken from the hopper at intervals. The digital control system controls a flow valve supplying material to the hopper. Two sensors are provided, one to detect when the level L1 is reached by the material in the hopper, and one to detect when level L2 is reached by the material. Each sensor generates a signal when the material has passed the level set by the sensor. This signal would typically be a voltage, say 5 volts, to indicate that the material has passed the sensor, and perhaps 0 volts to indicate that the material has not passed the sensor. Hence the input signals have two values, one to indicate

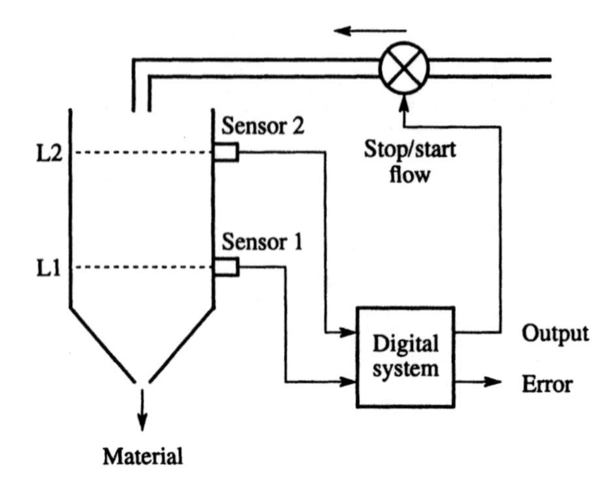

Figure 1.1 Controlling flow of material into a hopper

the presence of material and one to indicate the absence of material. The flow control valve will also be controlled by a two-state signal, ON and OFF in this case. For example, a 50 volts signal might turn the valve on and 0 volts might turn it off. The flow valve is to be turned on when the material is below L1 and kept on until L2 is reached when the valve is turned off. Again the valve is kept off until the material drops to L1. The digital system must generate a signal which implements this "algorithm". The signal has two values, one for ON and one for OFF.

Two levels are sensed in the system so that material can be allowed into the hopper for a reasonable period. Maintaining a single level by turning the valve on when below the level and turning the valve off when above the level would create a situation where the valve would be continually turned on and off.

Logic signals and functions

In subsequent chapters, we will study how to design a digital system such as that shown in Figure 1.1. The important point to observe here is that the input and output signals have two values, and two voltages will be used, one for each value. The digital system itself might use 5 volts and 0 volts to represent the two values. Rather than refer to specific voltages, we will refer to two-valued *logic* signals. The digital system will be designed with components that can accept two-valued logic signals and generate two-valued logic signals. The two values in our example correspond to ON and OFF. Sometimes the terms TRUE and FALSE are used, e.g. the signal from sensor 1 is TRUE when the material level is above level 1, and FALSE that the material level is below level 1. TRUE corresponds to ON and FALSE corresponds to OFF. The numerical values 1 and 0 are usually used in place on TRUE and FALSE in logic design; 1 corresponds to TRUE and 0 corresponds to FALSE. Table 1.1 relates the situation here and the corresponding "logic" signals.

Table 1.1 *Logic signals*

Control situation	*Logic voltage*	*Switch*	*Logic value*	
Below level	0 V	OFF	FALSE	0
Above level	5 V	ON	TRUE	1

Positive and negative logic representation

The actual voltages used in a system are typically 0 volts and 5 volts, but can be lower notably for the internal gates of complex components such as microprocessors. Given two voltages, say 0 volts and 5 volts, we must decide which voltage will be used to represent which logic value. For example, we could use 0 volts to represent a logic 1 and 5 volts to represent a logic 0. Whatever decision is made concerning the voltages, the same voltages are used throughout the system. Usually the higher voltage is chosen to represent a logic 1 and the lower voltage used to represent a logic 0 (e.g. 5 volts = 1 and 0 volts = 0). This is known as *positive logic representation.* An alternative is to use the higher voltage to represent a logic 0 and the lower voltage to represent a logic 1 which is known as *negative logic representation.* It is possible to use both positive and negative logic representations in a single logic system (while still using only two voltages). Using both representations in the same system is know as *mixed logic representation.* Then a special notation is needed for naming logic signals (see Fletcher, 1980 or Tinder, 1991).

Logic functions

To be able to develop the "algorithm" or logic function for a digital system, logic signals must be represented by names just as in algebra. For example, the two inputs in Figure 1.1 might be called S_1 and S_2 (for sensor 1 and sensor 2 respectively). The output might be called Z. In Figure 1.1, a second output signal is shown, a signal indicating that an error has occurred. An example of an error condition would be sensor 1 showing the material below level L1 and sensor 2 showing the material above level L2, which is of course impossible if all the components are working. Let the error output be E. The "algorithm" for the error signal is:

$$E = 1 \text{ if } S_1 = 0 \text{ and } S_2 = 1$$

This logic function can be written in a two-valued algebra called *Boolean algebra*:

$$E = \overline{S}_1 \cdot S_2$$

In this notation, $\overline{S}_1$ will be a 0 when S_1 is a 1 and $\overline{S}_1$ will be a 1 when S_1 is a 0. The dot ($\cdot$) corresponds to "and".

There could be more than one error condition, say E_1, E_2 and E_3, each of which indicates an error. If E_1 is a 1 or E_2 is a 1 or E_3 is a 1, we might want a general alarm to be sounded. The algorithm for this alarm is:

Alarm = 1 if $E_1 = 1$ or $E_2 = 1$ or $E_3 = 1$.

The logic function for the alarm is written in Boolean Algebra as:

$$\text{Alarm} = E_1 + E_2 + E_3$$

Here + corresponds to "or". In fact, we have now introduced the three fundamental operations of Boolean algebra, AND (·), OR (+) and NOT ($\bar{}$). From these operations, we can create any logical function. The full range of this algebra will be developed in Chapter 2. For the controller application, we need a logic circuit having two inputs, S_1 and S_2 and a single output, E, which will be a 1 if $S_1 = 0$ and $S_2 = 1$, otherwise $E = 0$. This requires a so-called *combinational logic circuit* because the output value will depend upon certain combinations of input values. The alarm function is also implemented with a combinational logic circuit.

The function for the output, Z, is slightly more complex as it requires the circuit to memorize logic values. Z will change to a 1 when $S_1 = 0$ but does not change back to a 0 when $S_1 = 1$ but only when $S_2 = 1$. This is a so-called *sequential logic circuit*, because generally the outputs of such circuits will follow sequences. Designing such logic circuits is addressed first in Chapter 4.

Calculations

More elaborate control applications might use sensors which take measurements. There might be a sensor which measures temperature. The material in our hopper might need to be maintained at a specific temperature by the use of heating elements. For example, the heater might be turned on when the temperature is below 40°C and kept on until a temperature of 45°C is reached. The heater is then turned off until the temperature has fallen to below 40°C. This is a very similar algorithm to that used to control the hopper, except now we need to handle numbers rather than simply ON/OFF values. In this and other applications, there might be many measurements and complex calculations necessary based upon the measurements. For example, a modern automobile engine is usually controlled internally by a digital system. Here the control of actions within the engine for optimum performance are dependent upon several variables including the current speed, the load, temperature, etc. Again, we now must operate upon numbers, not just two-valued logic variables.

Digital computer

Performing calculations in a semi-automatic way has long been a desire to relieve the burden of manual calculations. A mechanical calculator was devised by Babbage in the 1850s, and electronic versions began to be developed during the second world war, for example to calculate paths of projectiles. Babbage conceived the idea of a universal programmable calculator, the electronic version we now call a "digital computer". The digital computer represents one of the most complex digital systems, a system which manipulates numbers and calculates values from these numbers using a program of steps stored in the memory of the computer. Most complex

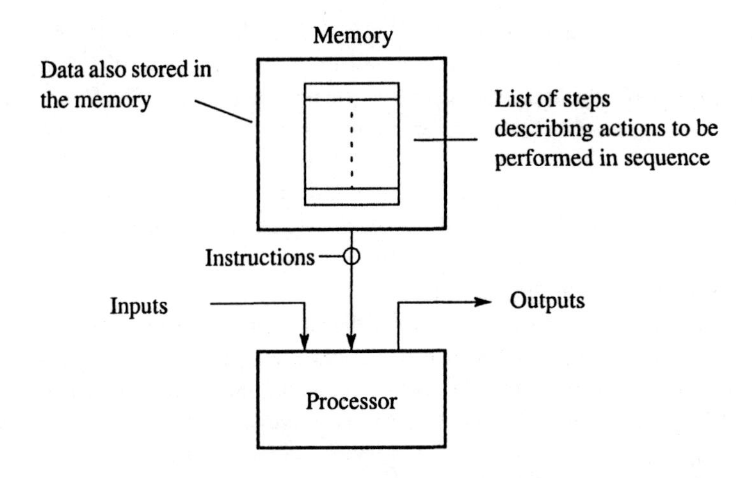

Figure 1.2 Programmable "computer"

control applications, such as for an automobile, are now performed by digital computers rather than special-purpose controllers.

Figure 1.2 shows the general arrangement of a digital computer. The processor is capable of performing various arithmetic operations on numbers and making decisions of what to do based upon the results of calculations. The actions to be performed are encoded in a numerical form and stored in a memory as instructions. The numerical form is "binary" which is described later. The encoded instructions are presented to a processor for execution in a sequence. Of course, the list of instructions to be executed (the *program*) must be developed by a human *programmer.* The processor must be precisely instructed on what calculations to do and on what basis the decisions are to be made, which is done by the formulation of the program.

1.2 Representing numbers in a digital system

1.2.1 Binary numbers

Whether our digital system is a fully programmable computer or a specialized control system manipulating numbers, we need a way to represent numbers with voltages. Suppose we were to use decimal numbers with one signal for each digit. Each signal would need to have ten different voltages, one for each possible value of each digit. For example, the 0 digit could be represented by 0 volts, the 1 digit by 1 volt, the 2 digit by 2 volts and so on. Then electronic circuits would have to be designed which can accept ten different voltages and produce ten different voltages, one for each digit of the number.

The decimal number system is a *positional* number system. It uses digits multiplied by powers of 10 that depend upon the position of the digit. For example, the decimal number $235 = 2 \times 10^2 + 3 \times 10^1 + 5 \times 10^0$. The 10 here is the *base* of the number system and results in requiring ten different digits values, 0, 1, 2, 3, 4, 5, 6, 7, 8 and 9. Ten was chosen probably because we have ten fingers. However any fixed value could be used for the base. In general an *n*-digit number in the base *b* is given by:

$$a_{n-1}a_{n-2} \ldots a_i \ldots a_2a_1a_0 = a_{n-1}b^{n-1} + a_{n-2}b^{n-2} \ldots + a_ib^i \ldots + a_2b^2 + a_1b^1 + a_0b^0$$

where a_i is the *i*th digit of the number.

If we were to use the base of 9, nine different values would be needed for the digits, 0, 1, 2, 3, 4, 5, 6, 7 and 8. A number must use these digits only. Hence the number 235 in this system represents the number computed (in decimal) as $2 \times 9^2 + 3 \times 9^1 + 5 \times 9^0$ (which equals 194 in decimal). If we were to use the base of 8, the digits are 0, 1, 2, 3, 4, 5, 6 and 7. The number 235 in this number system (the *octal* number system) would be $2 \times 8^2 + 3 \times 8^1 + 5 \times 8^0$ (which equals 157 in decimal). For clarification of the base being used, we might write a decimal number as 235_{10}, a number using the base 9 as 235_9 and a number using the base 8 as 235_8. Continuing to reduce the size of the base, the number of different digits reduces. With base of three (the *ternary* number system), we get three digits, 0, 1 and 2. With the base of two (the *binary* number system), the digits are 0 and 1. The digits of a binary number are called *bits* (*bi*nary digi*ts*). A bit can be 0 or 1. An example of a binary number is 110001_2 which equals $1 \times 2^5 + 1 \times 2^4 + 0 \times 2^3 + 0 \times 2^2 + 0 \times 2^1 + 1 \times 2^0 = 49$ in decimal.

SELF-ASSESSMENT QUESTION 1.1

What is the binary number 1101_2 in decimal?

Numbers can be a fraction or have a fractional part. The digits of the fractional part use negative powers of the base. In decimal $0.12_{10} = 1 \times 10^{-1} + 2 \times 10^{-2}$. In binary $0.101_2 = 1 \times 2^{-1} + 0 \times 2^{-2} + 1 \times 2^{-3}$.

SELF-ASSESSMENT QUESTION 1.2

What is the binary number 1101.101_2 in decimal?

We can see immediately that the binary number system fits in well with a system with ON/OFF control signals. More importantly though, it also has several advantages over other number systems, which come about because there are only two digits used, 0 or 1, in all numbers. Only two different voltages are needed for each digit, one for a 0 and one for a 1. For example, 0 volts could represent a digit when it is 0 and 5 volts could represent a digit when it is 1. This greatly simplifies dealing with engineering factors and in designing electronic circuits. Using two voltages also makes it easier to handle the variations in these voltages that can occur in practice due to variations in circuit components, external electromagnetic radiation and electrical noise.

We often need ways in a digital system to remember digit values. In such cases, one memory "cell" will be required for each digit. In that context, binary representation is advantageous because several types of memory naturally suit remembering two values. Memory based upon magnetic media (such as floppy disks and hard disks) can represent the two values using the two directions of magnetism through or across a medium. A part of the medium is assigned for one bit, and this can exist in one of two states corresponding to the two directions of magnetism.[1] One type of memory based upon semiconductor technology uses charge on a capacitor. The presence of charge indicates a 1 (say) and the absence of charge indicates a 0 (say). Optical systems could use the presence of light to indicate a 1 and the absence of light to indicate a 0. A mechanical switch or relay can be in two positions, on and off. The advantage of needing only two states was also taken in the past by using holes in paper tape and cards. The presence of a hole in a particular position represented a 1 and the absence of a hole represented a 0. In fact, the presence or absence of "something" is quite universally applicable to represent 1 and 0.

For all the reasons described, the binary number system has been adopted universally in digital systems.

SELF-ASSESSMENT QUESTION 1.3

Can you think of any other possibilities for representing binary digits?

The advantages of binary do not come without a "cost". As the size of the base is reduced, the number of different digits needed is reduced, but potentially increasing the number of digits required, depending upon the size of the number. Table 1.2 shows how this happens with reducing bases, from 10 to 2 (2 being the lowest possible base). We see that the number of digits to represent a particular number can increase with decreasing base. For example, the number 15 in decimal requires two digits when the base is 10, but requires three digits when the base is 3 and four digits when the base is 2. The actual number of digits required will depend upon the number. In general, if a base of b is used, there needs to be b different values for the digits and $\lceil \log_b x \rceil$ digits, where x is the number. The base of 2 can require typically three times more digits than the base of 10. Hence a basic cost of using binary is that three times the wires and circuits will be needed, as one wire/circuit is generally used for each digit of the number. The circuits themselves of course will be much simpler, and practical.

Binary numbers within the digital system or computer are normally held with a fixed number of bits. A group of eight bits is called a *byte*, a convenient size for holding the binary codes for alphanumeric data, see Section 1.5. For numbers, a larger group of bits is often used. A computer, for example, would often allocate thirty-two bits for numbers. Leading zeros are introduced to (positive) binary numbers when necessary. (Negative numbers will be considered later.)

---

[1] The two directions may not directly correspond to the two values; for example a change of direction may represent a 1 and no change of direction may represent a 0. For a description of digital magnetic recording techniques, see Wilkinson and Horrocks, 1987.

Table 1.2 *Numbers with different bases*

Base								
2	*3*	*4*	*5*	*6*	*7*	*8*	*9*	*10*
0	0	0	0	0	0	0	0	0
1	1	1	1	1	1	1	1	1
10	2	2	2	2	2	2	2	2
11	10	3	3	3	3	3	3	3
100	11	10	4	4	4	4	4	4
101	12	11	10	5	5	5	5	5
110	20	12	11	10	6	6	6	6
111	21	13	12	11	10	7	7	7
1000	22	20	13	12	11	10	8	8
1001	100	21	14	13	12	11	10	9
1010	101	22	20	14	13	12	11	10
1011	102	23	21	15	14	13	12	11
1100	110	30	22	20	15	14	13	12
1101	111	31	23	21	16	15	14	13
1110	112	32	24	22	20	16	15	14
1111	113	33	30	23	21	17	16	15

SELF-ASSESSMENT QUESTION 1.4

What is the largest binary number that can be held in a byte, given only positive numbers are represented?

1.2.2 Number conversion

The reasons have been established for using the binary number system within the digital system. However, we much prefer to use the decimal number system (even if we do not count on our fingers). Numeric data is likely to be entered using the decimal representation. Numerical results are preferred in decimal. Hence, there needs to be some way of converting from binary to decimal and vice versa if only a method that the computer can use.

Small decimal numbers can quite easily be converted to binary in an informal way by considering the values of powers of 2:

Decimal	64	32	16	8	4	2	1
	2^6	2^5	2^4	2^3	2^2	2^1	2^0
Binary	1000000	0100000	0010000	0001000	0000100	0000010	0000001

If we want to convert the decimal number 25 into binary, the powers of 2 that add up to the number would be found, starting with largest power of 2 less than 25, i.e.:

$$25 = 16 + 8 + 1 = 10000 + 01000 + 00001 = 11001$$

A formal conversion method can be derived from the definition of the "positional" number system (see Wilkinson, 1992, for full details).

SELF-ASSESSMENT QUESTION 1.5

What is the decimal number 234_{10} in binary?

An informal conversion from binary to decimal numbers is to simply add the powers of 2 as necessary to reach the number starting with the largest power of 2 less than the number and working towards the smallest power of 2 (zero), e.g.:

$$11001 = 16 + 8 + 1$$

Again a formal algorithm can be derived which starts with the definition of the positional number system (see Wilkinson, 1992, for full details).

1.2.3 Hexadecimal and octal numbers

The positional number system using the base 16 is called the *hexadecimal* number system. The hexadecimal number system has sixteen different digits, 0, 1, 2, 3, 4, 5, 6, 7, 8, 9 for the first ten values, 0 to 9. The remaining values are 10, 11, 12, 13, 14, 15. The letters of the alphabet are used for these values, i.e. A is used for 10, B for 11, C for 12, D for 13, E for 14 and F for 15. A hexadecimal digit is often referred to as a "hex" digit. Table 1.3 shows the sixteen hex digits and their equivalents in binary.

The hexadecimal number system has a special part to play here as there is a very simple way of converting between binary numbers and hexadecimal numbers. Because the base of hexadecimal ($16 = 2^4$) is the fourth power of the base of binary (2), a group of four binary bits corresponds to one hexadecimal digit. To convert a binary number into hexadecimal number is a simple matter of dividing the binary number into groups of four digits and converting each group into one hexadecimal digit. For example, to convert the binary number 10100111_2 into hexadecimal we have:

$$1010\ 0111_2 = A7_{16}$$

(1010_2 = 10 in decimal or A in hexadecimal.) The number of digits in the hexadeci-

Table 1.3 *Hexadecimal digits*

Decimal	*Binary*	*Hex*
0	0000	0
1	0001	1
2	0010	2
3	0011	3
4	0100	4
5	0101	5
6	0110	6
7	0111	7
8	1000	8
9	1001	9
10	1010	A
11	1011	B
12	1100	C
13	1101	D
14	1110	E
15	1111	F

mal number will be one fourth the number of digits in the binary number, assuming that the number of digits in the binary number is a multiple of four. Hence binary numbers can be written in hexadecimal with ease and with far fewer digits, while still being able to recognize easily the pattern of 0's and 1's in the binary number. This leads to hexadecimal representation being widely used in place of binary in documentation.

The method of converting between hexadecimal and binary can be derived from the original definition of the position number system. To take another example, converting $1AC_{16}$, we get:

$$1AC_{16} = (0001_2) \times 16^2 + (1010_2) \times 16^1 + (1100_2) \times 16^0$$

which equals:

$$(0 \times 2^3 + 0 \times 2^2 + 0 \times 2^1 + 1 \times 2^0) \times 16^2 + (1 \times 2^3 + 0 \times 2^2 + 1 \times 2^1 + 0 \times 2^0) \times 16^1$$
$$+ (1 \times 2^3 + 1 \times 2^2 + 0 \times 2^1 + 0 \times 2^0) \times 16^0$$

Because $16^2 = 2^8$, $16^1 = 2^4$, $16^0 = 2^0$, we get:

$$(0 \times 2^{11} + 0 \times 2^{10} + 0 \times 2^{9} + 1 \times 2^{8}) + (1 \times 2^{7} + 0 \times 2^{6} + 1 \times 2^{5} + 0 \times 2^{4}) + (1 \times 2^{3} + 1 \times 2^{2} + 0 \times 2^{1} + 0 \times 2^{0})$$

which equals 000110101100_2

SELF-ASSESSMENT QUESTION 1.6

Convert the hexadecimal number $3D_{16}$ into binary.

SELF-ASSESSMENT QUESTION 1.7

Convert the binary number 100001010100010_2 into hexadecimal.

The octal number system also has a simple conversion method. Because the octal base, 8, is the third power of the binary base, three binary bits correspond to one octal digit. For example, the octal number 345_8 in binary is given by:

$$345_8 = (011\ 100\ 101)_2$$

However octal is not as useful as hexadecimal as most systems have binary words which are multiples of bytes. One byte will convert directly to two hex digits. Octal was useful in older computer systems employing 24-bit numbers as then eight octal digits would convert to one 24-bit binary word.

1.3 Performing arithmetic with binary numbers

1.3.1 Addition

Clearly with binary numbers being used inside our digital system, we will need ways to perform arithmetic upon binary numbers. Fortunately arithmetic using binary numbers is particularly simple. We learn the tables for decimal addition, subtraction, and multiplication, which requires every combination of the ten digits in decimal. In binary, similar tables can be constructed, but there are only four combinations of two digits, 0 and 0, 0 and 1, 1 and 0, and 1 and 1. For addition, $0 + 0 = 0$, $0 + 1 = 1$, $1 + 0 = 1$, and $1 + 1 = 10$. (Compare with the equivalent in decimal: $0 + 0 = 0$, $0 + 1 = 1$, $0 + 2 = 2$, $\cdots 9 + 9 = 18$.) Binary addition in tabular form is shown in Table 1.4.

Table 1.4 *Adding two binary digits together*

A	*B*	*SUM*		*Decimal*
0	0	0	0	0
0	1	0	1	1
1	0	0	1	1
1	1	1	0	2

Suppose now two binary numbers, each composed of several bits, are to be added together. First the least significant pair of digits would be added together as we would in decimal. In decimal, if the result was greater than 9, a "carry" is generated which is added to the next pair of digits (e.g. $7+8=5$ and a carry of 1) and so on with subsequent pairs of digits. In binary, if the number is greater than 1, a carry is generated to be added to the next pair of digits (e.g. $1+1=0$ and a carry of 1). Just as in decimal, we will now need to add three digits, the two digits of the two numbers together with a carry digit. There are eight combinations of three binary digits as shown in Table 1.5:

Table 1.5 *Adding three binary digits together*

A	B	C_{in}	C_{out}	*SUM*	*Decimal*
0	0	0	0	0	0
0	0	1	0	1	1
0	1	0	0	1	1
0	1	1	1	0	2
1	0	0	0	1	1
1	0	1	1	0	2
1	1	0	1	0	2
1	1	1	1	1	3

To take a comparison of adding in decimal and adding in binary, suppose the numbers 22_{10} and 28_{10} are to be added. It would look like:

Decimal	Binary
2 2	1 0 1 1 0
$2{}_{1}8$	${}_{1}1{}_{1}1{}_{1}1\,0\,0$
5 0	1 1 0 0 1 0

The rules are essentially the same whether decimal or binary (or any other base) is used. Pairs of digits are added together, starting with the least significant digits. When a result digit is equal to or greater than b (where b is the base), a carry is generated which is added to the next pair of digits. For decimal, results greater than 9 produce a carry; for binary, results greater than 1 produce a carry.

SELF-ASSESSMENT QUESTION 1.8

Add 81_{10} and 63_{10} in binary. Confirm the binary answer by converting it back to decimal.

To implement the paper and pencil approach directly, a circuit will be needed for the first stage which takes two binary digits and creates a sum digit and a carry digit for the next stage. For each of the subsequent stages, we need a circuit which takes three digits, the two digits of the number together with the carry from the previous stage, to create two digits, a sum digit and a carry digit for the next stage. We shall design such logic circuits in Chapter 3.

1.3.2 Negative numbers and subtraction

Binary subtraction could use a similar approach to the decimal approach of "borrowing" from the next column of digits. In decimal, we have to borrow a digit from the next column if the *subtrahend* (the digit to be subtracted) is greater than the *minuend* (the digit that is being subtracted from). For example, to subtract 7 from 6, we have to borrow 10 and subtract 7 from 16. However for implementation reasons (reasons that will make the construction of the circuits simpler), subtraction is usually done by the "addition of complements". Such methods can be applied to any number system and revolve around the complement of a number being used to represent the negative of the number. In decimal, the 10's complement of a decimal number, N, is defined as:

$$10^n - N$$

where there are n digits in the number. Suppose there were three digits. Then the 10's complement of the decimal number 235 would be $1000 - 235 = 765$.

Similarly, the 2's complement of a binary number, N, is given by:

$$2^n - N$$

where there are n digits in the number. Suppose there were three digits. Then the 2's complement of the binary number 011 (3) would be $1000 - 011 = 101$.

The 2's complement representation will be the representation for negative numbers. For example, the representation of –3 in binary will be 101. The representation of positive numbers will be as normal, i.e. $+3 = 011$. It is necessary to provide sufficient digits to accommodate both positive and negative numbers, and this means that at least one leading zero will be used for positive numbers. Negative numbers will always start with a leading 1. Table 1.6 shows how the numbers from –8 through to +7 are represented using four digits. Notice that positive numbers are unchanged from their "pure" binary representation. Also notice that the range of numbers provides for one more negative number than positive number, –8. This number does not have a positive equivalent, and is an exception to the previous definition, i.e. to obtain –8 we cannot compute $2^3 - 8$. (This is zero.) However the representation is consistent. For example, if we add 1 to –8, we get –7 ($1000 + 0001 = 1001$).

Usually a fixed number of digits are provided in a system for numbers, say thirty-two digits (bits). Using the 2's complement representation, numbers can have values between -2^n through zero to $+2^n - 1$ when there are n digits. To extend a positive

Table 1.6 *Two's complement representation*

Decimal	*Binary*
+7	0111
+6	0110
+5	0101
+4	0100
+3	0011
+2	0010
+1	0001
0	0000
–1	1111
–2	1110
–3	1101
–4	1100
–5	1011
–6	1010
–7	1001
–8	1000

number to a given number of digits, we simply insert leading zeros. To extend a negative number we insert leading ones. For example –6 = 1010 with four digits, 11111010 with eight digits.

Negative numbers are represented within most digital systems and computers in 2's complement because it simplifies addition and subtraction. Let us explore the complement representation in binary and its use for addition and subtraction. (The same results apply to decimal complements, should we wish to use the method for decimal subtraction.)

Subtraction

To perform subtraction of Y from X, the (2's) complement of Y is formed to create $-Y$. Then we add $-Y$ to X. In doing this, we perform:

$$X - Y = X + (-Y) = X + (2^n - Y) = X - Y + 2^n$$

If X is greater than Y, $X - Y$ is positive. The term 2^n is a number given by 1 and n zeros (1000 … 000). If we truncate the number at n digits, the 2^n term will be ignored and we get the correct positive answer. If $X - Y$ is negative, the correct negative answer in the 2's complement representation is produced naturally, i.e. $2^n - (Y - X)$.

To establish completely that the representation is suitable, we should prove the result for every combination of positive and negative numbers. If Y is already negative, we will negate a negative Y to get a positive Y, i.e.:

$$-(-Y) = 2^n - (2^n - Y) = 2^n - 2^n + Y = Y$$

It now follows that all combinations will work.

EXAMPLES

(a) *X and Y positive, and $X > Y$. The answer is positive.*

Suppose $X = 0110$ (6) and $Y = 0011$ (3). Then $-Y = 1101$ (−3), and 6 − 3 is given by:

Decimal	Binary
6	0 1 1 0
−3	+ $_{1}1_{1}1\,0\,1$
3	0 0 1 1

Notice a final carry is generated corresponding to the 2^n term, but ignored.

(b) *X and Y positive, and $X < Y$. The answer is negative.*

Suppose $X = 0011$ (3) and $Y = 0110$ (6). Then $-Y = 1010$ (−6), and 3 − 6 is given by:

Decimal	Binary
3	0 0 1 1
−6	+ $1_{1}0\,1\,0$
−3	1 1 0 1

A final carry is not generated in this case, and the answer is a correct negative representation of 3.

(c) *X positive and Y negative. The answer is always positive.*

Suppose $X = 0011$ (3) and $Y = 1101$ (−3). Then $-Y = 0011$ (+3), and 3 − (−3) is given by:

Decimal	Binary
3	0 0 1 1
−(−3)	+ $0\,0_{1}1_{1}1$
6	0 1 1 0

(d) *X negative and Y positive. The answer is always negative.*

Suppose $X = 1101$ (–3) and $Y = 0011$ (3). Then $-Y = 1101$ (–3), and –3 – 3 is given by:

Decimal	Binary
–3	1 1 0 1
–3	$+\ {}_1 1\, {}_1 1\ 0\, {}_1 1$
–6	1 0 1 0

A final carry is generated corresponding to the 2^n term but is ignored.

Clearly in all cases, we are still performing subtraction to create $-Y$ from $+Y$ (subtracting Y from 2^n). If this subtraction had to be done by a normal means of using borrows, we have not eliminated the subtraction process. Fortunately the 2's complement result can be obtained without subtracting Y from 2^n. The number 2^n is 100 … 00 (1 and n zeros). Therefore $2^n - 1$ is given by 011 … 11. If we subtract any number from 011 … 11, we invert the digits of the number (change all 1's to 0's and all 0's to 1's). For example, subtracting 0110 from 1111, we get 1001, ignoring the $(n-1)$st digit as usual. It is a simple matter to invert the digits. But we want to subtract from 2^n, not $2^n - 1$. We must now add one to the result, i.e. $2^n - X = (2^n - 1) - X + 1$. Therefore we have simple method to create the 2's complement of any number: "invert the digits and add 1". We shall see in Chapter 3 that this method can be easily and efficiently implemented in hardware. Note that "invert the digits and add 1" is only a method. It is not the definition of the 2's complement number – that is given by $2^n - N$.

SELF-ASSESSMENT QUESTION 1.9

Convert the binary number 0101010 into its 2's complement negative using the method "invert the digits and add 1".

SELF-ASSESSMENT QUESTION 1.10

Convert the answer for Self-assessment question 1.9 back to a positive number using the same method.

An alternative algorithm for converting a number to its 2's complement form is to copy the digits from right to left until reaching the first 1, copy this digit and then invert all remaining digits to the left. (Proof: Tutorial question 1.10.)

The 2's complement representation can also be viewed in different ways. We can view the digits of the number as having the weighting:

$$-2^{n-1}\ \ 2^{n-2}\ \ 2^{n-3}\ \ \ldots\ \ 2^1\ \ 2^0$$

For example the number –6 is:

$$\begin{array}{ccccl} -2^3 & 2^2 & 2^1 & 2^0 & \\ 1 & 0 & 1 & 0 & = -2^3 + 2^1 = -6 \end{array}$$

This comes about directly from the definition of a 2's complement number (Tutorial question 1.11).

SELF-ASSESSMENT QUESTION 1.11

Convert the binary number 0101010 into its 2's complement negative using the rule "copy the digits from right to left until the first 1, copy this digit and then invert all remaining digits to the left". Show that the result is correct.

Finally notice that the 2's complement approach eliminates any special treatment of numbers in subtraction when the subtrahend is larger than the minuend, as is necessary in paper-and-pencil subtraction of decimal numbers.

SELF-ASSESSMENT QUESTION 1.12

Subtract 63_{10} from 14_{10} in decimal. Repeat in binary.

One's complement

Inverting the digits of a binary number is called forming the (1's) complement which can be defined as:

$$-X = (2^n - 1) - Y$$

It turns out that arithmetic can also be performed using the 1's complement representation directly rather than the 2's complement representation, though in this case the final carry indicates that an extra 1 must be added to the result. This can be easily proved from the definition of a 1's complement representation (Tutorial question 1.12). Most computers use the 2's complement representation for negative numbers. The decimal version of binary 1's complement is the decimal 9's complement with the definition:

$$-X = (10^n - 1) - Y$$

1.3.3 Binary-coded decimal numbers

Some applications such as calculators and checkout tills involve simply processing decimal numbers and displaying decimal numbers. Decimal numbers are entered at a keypad and results are displayed in decimal. Clearly the decimal data can be converted into binary and binary results converted back into decimal for display. An alternative is to retain the decimal nature of the information within the system using a coding called *binary-coded decimal* (BCD). In BCD, each decimal digit of a number is represented by a 4-bit binary code. The 4-bit codes of each digit are simply concatenated. For example, the decimal number 2345_{10} would be represented in BCD as the code: 0010 0011 0100 0101.

Once in this representation, arithmetic addition of numbers can proceed as though the numbers are normal binary, with certain changes. Suppose we were to add two decimal numbers 2431_{10} and 5425_{10} as follows:

Decimal	BCD
2431	0010 0100 0011 0001
5425	0101 0100 0010 0101
7856	0111 1000 0101 0110

We can see that by adding the codes as though they were binary actually produced the correct answer in BCD. This example was carefully chosen so that each digit of the result was less than 10. If a digit of the result were greater than 10, the correct answer would not be obtained immediately. A so-called "correction factor" of 6 must be added. This is to jump over the six unused patterns in BCD that represent 10, 11, 12, 13 14 and 15, namely 1010, 1011, 1100, 1101, 1110, and 1111.

When the result digit is between 10 decimal and 15 decimal, simply looking at the four bits of the BCD digit will suffice, as in the example below:

	Decimal	BCD
	3 6	0011 0110
	2 5	0010 0101
	5 11	0101 1011
correction:	6	0110
	6 1	0110 0001

When the result is 16, 17 or 18, the four bits of the BCD digit will be 0000, 0001, or 0010 respectively. A carry will have been generated which will affect the next BCD digit. Hence simply looking at the four result bits will not suffice; the carry must be recognized as in the following example:

	Decimal	BCD
	3 8	0011 1000
	2 9	$0010_{1}1001$
	6 1	0110 0001
correction:	6	0110
	6 7	0110 0111

Computers designed for BCD applications have special BCD instructions which perform the correction automatically. Subtraction can be done by subtracting the correction factor when necessary. BCD was particularly suitable in the past when using early 4-bit single chip microprocessors, but less so now.

1.4 Representation of alphanumeric symbols

Apart from numbers, computers must have a way to represent letters of the alphabet and other symbols that can be entered at a keyboard or output on a display or to a printer. The term *alphanumeric* is used to describe these symbols which includes the decimal digits 0 to 9. A standard code for alphanumeric symbols has been defined in the American Standard Code for Information Interchange (ASCII), as given in Table 1.7. This table defines a 7-bit code, by reading the bits across the top, $b_7b_6b_5$ concatenated with the bits reading down the side, $b_4b_3b_2b_1$. For example, the ASCII code for the letter *B* is 1000010.

Table 1.7 *7-bit ASCII code*

$b_4b_3b_2b_1$	$b_7b_6b_5$	000	001	010	011	100	101	110	111
		0	1	2	3	4	5	6	7
0000	0	NUL	DLE	SP	0	@	P	‘	p
0001	1	SOH	DC1	!	1	A	Q	a	q
0010	2	STX	DC2	"	2	B	R	b	r
0011	3	ETX	DC3	#	3	C	S	c	s
0100	4	EOT	DC4	$	4	D	T	d	t
0101	5	ENQ	NAK	%	5	E	U	e	u
0110	6	ACK	SYN	&	6	F	V	f	v
0111	7	BEL	ETB	'	7	G	W	g	w
1000	8	BS	CAN	(	8	H	X	h	x
1001	9	HT	EM	)	9	I	Y	i	y
1010	10	LF	SUB	*	:	J	Z	j	z
1011	11	VT	ESC	+	;	K	[	k	{
1100	12	FF	FS	,	<	L	\	l	|
1101	13	CR	GS	-	=	M	]	m	}
1110	14	SO	RS	.	>	N	^	n	~
1111	15	SI	US	/	?	O	_	o	DEL

SELF-ASSESSMENT QUESTION 1.13

What is the ASCII code for the letter *w*?

SELF-ASSESSMENT QUESTION 1.14

How can the code for lower-case letters be changed to the code for upper-case letters, irrespective of the letter?

1.5 Digital logic example

In this final section, we shall look at a simple application of digital logic, as an introduction to a more detailed study in subsequent chapters. This application will introduce the two basic types of logic circuits, the *combinational logic circuit* and the *sequential logic circuit*. Each type will be studied in separate chapters.

A computer keyboard is arranged as a typewriter with rows of keys for the letters of the alphabet, numerals and punctuation and special symbols.[2] Usually the keys of the keyboard are connected electrically in a two-dimensional array with n columns and m rows (say) as shown in Figure 1.3. A logic 1 signal is applied to each of the rows in turn. When a key is pressed the signal is passed to the columns on which the key is placed. The column is identified and hence the key. (Additional components are needed to prevent erroneous conduction paths being made when several keys are pressed simultaneously.)

Suppose there are four rows. A 2-bit number is sufficient to identify the row. This

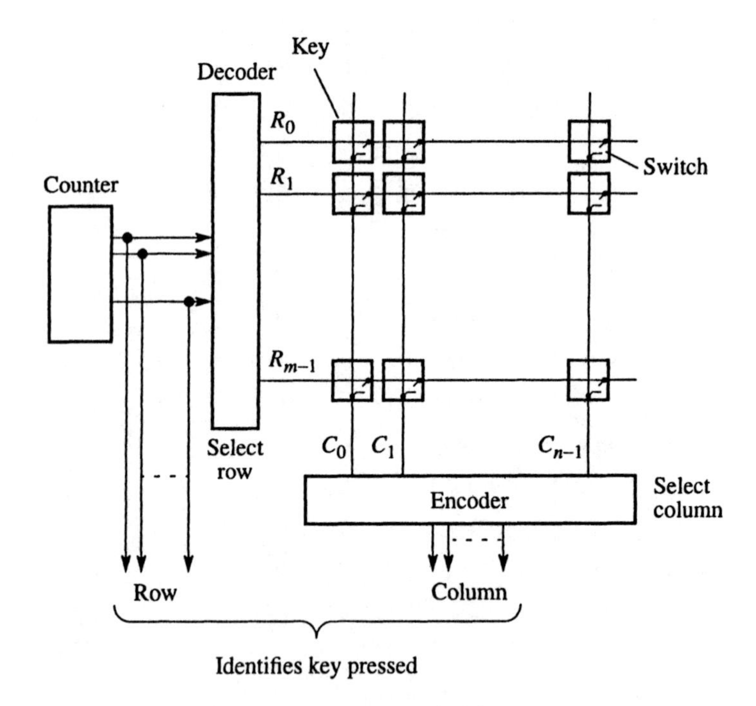

Figure 1.3 Keyboard decoder

[2] The actual layout of a typewriter keyboard was defined many years ago and purposely to slow a typist down because the old mechanical keyboards could not cope with the very high speed that could be achieved by a typist if the keys were laid out in the best way for a typist! The concocted layout has persisted to the present day though present-day keyboards can handle any possible typing speed.

2-bit number needs to be translated into a 4-bit pattern with a 1 in the position of the row selected, as shown in Table 1.8. The circuit to convert the 2-bit pattern to the 4-bit pattern defined in Table 1.8 is called a 2-line-to-4-line *decoder.* (Two input lines selecting one of four output lines.) We shall look at how to design decoders in Chapter 3.

Table 1.8 *Keyboard row select pattern*

Row	R_3 R_2 R_1 R_0
0 0	0 0 0 1
0 1	0 0 1 0
1 0	0 1 0 0
1 1	1 0 0 0

To sequence through each pattern, we could use a 2-bit *counter,* a logic circuit whose outputs follow the binary sequence 00, 01, 10, 11 repeatedly. How to design counters is described in Chapter 4.

Finally we need a logic circuit which accepts the signals from columns and generates a binary number which identifies the column having the logic signal present. Suppose there are sixteen columns. A 4-bit number is required and a 16-line-to-4-line *encoder.* This circuit actually performs the opposite function to a 4-line-to-16-line decoder. A 16-line-to-4-line encoder implements the function defined in Table 1.9. This is another logic circuit we shall design in Chapter 3. We see from Figure 1.3 that the result of the circuit configuration is a row address and a column address of the key pressed. Actually each symbol on the keyboard is assigned an ASCII code and a translation is necessary.

1.6 Summary

Computers are of course familiar since they are pervasive in society and everyone has been exposed to the uses of computers. The most important point in this chapter is that two-state signals are to be used in computers and other digital systems for both numerical and logical values.

The reader should appreciate the following:

- Signals having only two values are the most convenient and universally used inside a digital system.
- Two-valued signals are used for two basic purposes, first to represent ON/OFF control actions, and second to represent numbers using binary representation.
- Hexadecimal numbers are a convenient intermediate representation.

Table 1.9 *16-line-to-4 line encoder*

C_{15}	C_{14}	C_{13}	C_{12}	C_{11}	C_{10}	C_9	C_8	C_7	C_6	C_5	C_4	C_3	C_2	C_1	C_0	Output
0	0	0	0	0	0	0	0	0	0	0	0	0	0	0	1	0 0 0 0
0	0	0	0	0	0	0	0	0	0	0	0	0	0	1	0	0 0 0 1
0	0	0	0	0	0	0	0	0	0	0	0	0	1	0	0	0 0 1 0
0	0	0	0	0	0	0	0	0	0	0	0	1	0	0	0	0 0 1 1
0	0	0	0	0	0	0	0	0	0	0	1	0	0	0	0	0 1 0 0
0	0	0	0	0	0	0	0	0	0	1	0	0	0	0	0	0 1 0 1
0	0	0	0	0	0	0	0	0	1	0	0	0	0	0	0	0 1 1 0
0	0	0	0	0	0	0	0	1	0	0	0	0	0	0	0	0 1 1 1
0	0	0	0	0	0	0	1	0	0	0	0	0	0	0	0	1 0 0 0
0	0	0	0	0	0	1	0	0	0	0	0	0	0	0	0	1 0 0 1
0	0	0	0	0	1	0	0	0	0	0	0	0	0	0	0	1 0 1 0
0	0	0	0	1	0	0	0	0	0	0	0	0	0	0	0	1 0 1 1
0	0	0	1	0	0	0	0	0	0	0	0	0	0	0	0	1 1 0 0
0	0	1	0	0	0	0	0	0	0	0	0	0	0	0	0	1 1 0 1
0	1	0	0	0	0	0	0	0	0	0	0	0	0	0	0	1 1 1 0
1	0	0	0	0	0	0	0	0	0	0	0	0	0	0	0	1 1 1 1

- Binary addition is similar in concept to decimal addition.
- Subtraction is most conveniently done using a complement representation for negative numbers.

1.7 Tutorial questions

1.1 Write the first thirty numbers in base 11 and in base 12.

1.2 Turing (Hodges, 1983) used a number system with the base of 32 for an early computer design in 1951. Write the first thirty-three numbers in this number system. Why do you think he wanted to use this base?

1.3 Convert the following decimal numbers into binary:

(a) 56
(b) 102

1.4 Convert the following binary integers into decimal:

(a) 1011
(b) 11111

1.5 Convert the following hexadecimal numbers into binary:

(a) 77
(b) AA

1.6 Convert the following binary numbers into hexadecimal:

(a) 1010101010
(b) 1111111111

1.7 Convert the following negative decimal numbers into 2's complement binary numbers:

(a) −12
(b) −345

1.8 Convert the following 8-bit 2's complement negative binary numbers into decimal:

(a) 111111111
(b) 10010000

1.9 Convert the following decimal numbers into binary and perform the arithmetic operation(s) indicated, using 2's complement arithmetic where appropriate:

(a) $32 + 32$
(b) $67 + 132$
(c) $67 - 32$
(d) $-24 + 43$
(e) $-88 - 99$

1.10 Prove that a number can be converted to its 2's complement form by the following algorithm: "Copy the digits from right to left until reaching the first 1, copy this digit and then invert all remaining digits to the left."

1.11 Prove that the 2's complement representation can be obtained by weighting each digit as follows:

$$-2^{n-1}2^{n-2}2^{n-3} \ldots 2^{1}2^{0}$$

1.12 From the definition of a 1's complement number, prove that the representation can be used for subtraction if a 1 is added to the result when a final carry is generated (the so-called *end-around carry*).

1.13 Identify the keyboard's actions in the keyboard described in Section 1.6 that will cause erroneous conduction paths and erroneous operation (if additional components are not incorporated).

1.8 Suggested further reading

Fletcher, W. I., *An Engineering Approach to Digital Design*, Prentice Hall: Englewood Cliffs, New Jersey, 1980.

Hodges, A., *Alan Turing: The Enigma*, Simon and Schuster: New York, 1983. (This book is a fascinating biography of Alan Turing (background for Tutorial question 1.2) which should be compulsory reading for all Computer Engineering and Computer Science students!)

Karam, G. M. and Bryant, J. C., *Principles of Computer Systems*, Prentice Hall: Englewood Cliffs, New Jersey, 1992.

Page, E. S. and Wilson, L. B., *Information Representation and Manipulation in a Computer*, Cambridge University Press: Cambridge, England, 1973. (This is an old (but still relevant) book *solely* on information representation.)

Tinder, R. F., *Digital Engineering Design, A Modern Approach*, Prentice Hall: Englewood Cliffs, New Jersey, 1991.

Wilkinson B. and Horrocks, D., *Computer Peripherals, 2nd edition*, Hodder and Stoughton: London, 1987.

Wilkinson, B., *Digital System Design, 2nd edition*, Prentice Hall: London, 1992.

CHAPTER 2

Logic gates

Aims and objectives

The purpose of this chapter is to establish the fundamental components used in a digital system, namely logic gates. The various types of logic gates will be described here, and most systems employ a variety of gates. The fundamental mathematical underpinning, Boolean algebra, will be developed in this chapter.

2.1 Logic signals

First let us recap the essential ideas of logic signals as given in Chapter 1. Two-valued signals are used throughout a digital system to represent ON/OFF control actions and to represent the digits of binary numbers. A voltage level will represent each of the two logic values. We might picture a digital system as shown in Figure 2.1. The logic signals may change over time (see, for example, the control system in Chapter 1, Section 1.1). It was mentioned in Chapter 1 that +5 V could be used to represent a logic 1 and 0 V used to represent a logic 0. Using the higher voltage to represent a logic 1 and lower voltage to represent a logic 0 is positive logic representation. Using the higher voltage to represent a logic 0 and the lower voltage to represent a logic 1 is negative logic representation. Often only one representation is used within a system, normally positive logic representation.

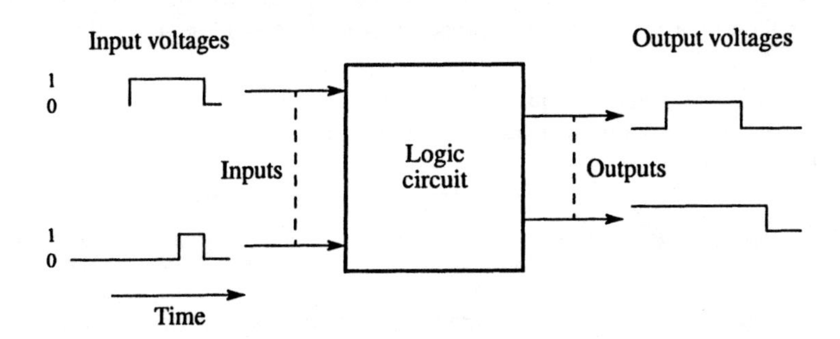

Figure 2.1 Logic circuit accepting logic voltages and generating logic voltages

Algebraic manipulation of two-valued variables

Our ultimate task is to design a logic circuit which will produce outputs according to certain application requirements. Applications have unique aspects and the most appropriate way to proceed is to design complex functions by using a collection of simple functions. Just as complex arithmetic expressions are created by using a few basic arithmetic operations (addition, subtraction, multiplication and division), logic functions can be specified by using simple logic operations. In fact, logic functions can be thought of as algebraic expressions where variables can take on only one of two values rather than an infinite number of values. Expressions with two-valued variables were developed by Boole in the 19th century. Then the motives were not to produce a digital system; it was to reason about the truth or otherwise of statements. However there is a direct relationship between the algebra of Boole (*Boolean algebra*) and "switching" circuits[1] (which was identified by Shannon in 1938).

2.2 Basic logic functions

There are three fundamental operations in Boolean algebra from which all logic functions can be developed, namely:

NOT
AND
OR

These and some other simple functions are implemented by circuits called *gates*. A gate accepts one or more logic signals and produces a logic output according to a basic logic function of the inputs. We have seen an informal introduction to the basic logic operations in Chapter 1. Now let us consider the operations more formally.

2.2.1 NOT gate

A logic variable can be either TRUE or FALSE (a 1 or a 0). We often need to be able to change a variable from one value to the value, i.e. from a 1 to a 0 or from a 0 to a 1. This is the fundamental NOT operation, which has the following definition:

> **Definition:** The NOT operation is applied to a single variable, A say, and produces the opposite logic value to A. If A is a 1, NOT A is a 0 and if A is a 0 NOT A is a 1.

The NOT operation on a variable A is written as $\bar{A}$. Some basic electronic circuits to

[1] The term *switching circuit* comes from the historical use of electromechanical switches to implement logic functions. We still use the term, especially when considering the internal design of the gates as described in Section 2.6.2.

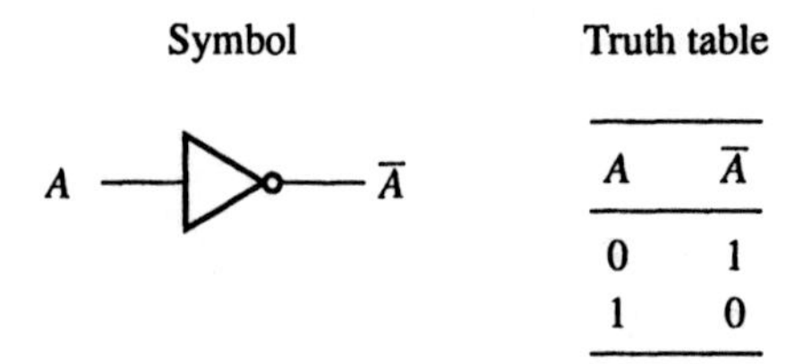

A	$\bar{A}$
0	1
1	0

Figure 2.2 NOT gate symbol and truth table

implement gates such as a gate to produce the NOT operation will be developed later. However, it is more convenient in logic design to hide the actual circuit implementations and draw circuits using symbols for the gates. Figure 2.2 shows the symbol for the NOT gate and its truth table. A truth table lists the output for each possible input value. Since in this case there is only one input, there are two possible input values and it is very reasonable to list the output for each value using a truth table. The NOT gate is sometimes called an *inverter*, and we talk of inverting a variable, or forming its *complement*. The bubble on the output of the symbol shows inversion. Bubbles are used freely in logic symbols to show inversion and, as we shall see, can be applied to both inputs and outputs.

SELF-ASSESSMENT QUESTION 2.1

What is the result of the NOT operation on $\bar{A}$?

2.2.2 AND gate

One of the fundamental operations we find in designing a digital system is detecting when two logic signals are both a 1. For example, we might design an adder which adds together two binary digits, say A and B to produce a sum digit and a carry digit. The carry digit is a 1 if both A and B are a 1 (see Chapter 1, Section 1.3.1). In control applications we can find many instances when some action is to take place if multiple conditions co-exist. The AND operation captures these situations with the following definition:

> **Definition:** The AND operation, operating on two variables A and B, produces a 1 if A is a 1 "and" B is a 1, otherwise the result is a 0.

We show the AND operation between two variables, A and B, by using the symbol ·, i.e. A "and" B is written as $A \cdot B$ so that $A \cdot B = 1$ if $A = 1$ and $B = 1$, otherwise $A \cdot B$ is a 0. Sometimes the · symbol is omitted just as the multiplication symbol is omitted between variables in ordinary algebra/arithmetic, i.e. $A \cdot B$ could be written as AB. The only possible problem with omitting the · symbol is if variable names could have more than one letter; then AB could be the name of a single variable. In

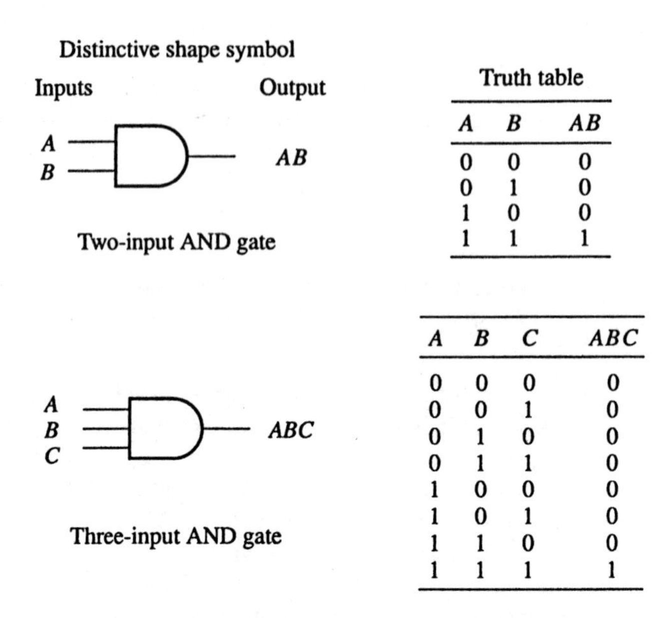

A	B	AB
0	0	0
0	1	0
1	0	0
1	1	1

A	B	C	ABC
0	0	0	0
0	0	1	0
0	1	0	0
0	1	1	0
1	0	0	0
1	0	1	0
1	1	0	0
1	1	1	1

Figure 2.3 AND gate symbol and truth table

fact, it may be preferable to using "multi-letter" variable names which are more meaningful that single letters, for example Motor_On to cause a motor to turn on. However because the $\cdot$ symbol is almost universally omitted, we shall omit it unless the operation needs to be clarified.

Just as ordinary multiplication, the AND operation can be applied to more than two variables. We could write the AND operation between several variables, i.e. A "and" B "and" C, which is written as $A{\cdot}B{\cdot}C$, or simply ABC. The result is a 1 if A is a 1 and B is a 1 and C is a 1, otherwise the result is a 0. There is only one instance that ABC is 1, when all the variables are a 1. Any number of variables could be connected with AND operators, and again the result is a 1 if all the variables are a 1 otherwise the result is a 0.

A specific number of inputs must be chosen for the AND gate. The number of inputs could be two, three, four, or more. Two-input and three-input AND gate symbols and corresponding truth tables are shown in Figure 2.3. The symbols shown are called *distinctive shape* symbols because they are designed to be immediately recognizable.

SELF-ASSESSMENT QUESTION 2.2

What is the result of $A{\cdot}A$?

2.2.3 OR gate

A common requirement is to show if at least one event is present as indicated by the associated variable being a 1. The example given in Chapter 1 was to detect one of several error conditions being present. Such situations can be captured in the OR operation, the final basic operation in Boolean algebra, which has the following definition:

> **Definition:** The OR operation applied to two variables, A and B, results in a 1 if $A = 1$ "or" $B = 1$, or both A and B are 1, otherwise the result is a 0 (i.e. when A and B are both a 0).

The symbol for the OR operation is +. Therefore we would show A "or" B as $A + B$. The same symbol is used as arithmetic addition, but there should be no ambiguity because arithmetic addition would not normally appear in a Boolean expression.

As with Boolean AND, Boolean OR can be applied to more than two variables. We would write the operations between the variables, i.e. A "or" B "or" C as $A + B + C$. The result is a 1 if any of A, B or C is a 1, or if any combination of A, B or C is a 1. There is only one instance of the function being a 0, when all of A, B and C are a 0. Two-input and three-input OR gate symbols and truth tables are shown in Figure 2.4.

SELF-ASSESSMENT QUESTION 2.3
What is the result of $A + A$?

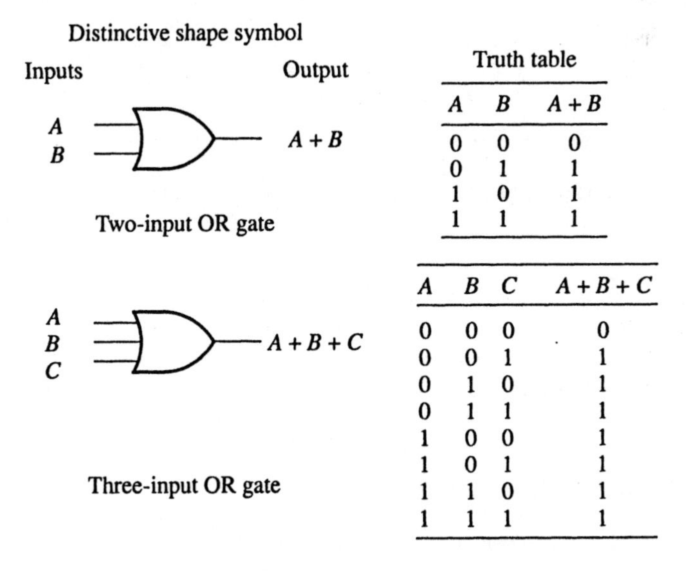

A	B	$A+B$
0	0	0
0	1	1
1	0	1
1	1	1

A	B	C	$A+B+C$
0	0	0	0
0	0	1	1
0	1	0	1
0	1	1	1
1	0	0	1
1	0	1	1
1	1	0	1
1	1	1	1

Figure 2.4 OR gate and truth table

2.3 Boolean relationships

2.3.1 Identities and their use

One question was posed after each of the sections on NOT, AND and OR gates. In fact, the answers are part of a set of identities, namely:

$$A \cdot 1 = A$$
$$A \cdot 0 = 0$$
$$A \cdot A = A$$
$$A + 1 = 1$$
$$A + 0 = A$$
$$A + A = A$$
$$A \cdot \bar{A} = 0$$
$$A + \bar{A} = 1$$
$$\bar{\bar{A}} = A$$

These identities are referred to as the basic Boolean identities. They can be easily proved by simply listing all possible values of the variables in a truth table as shown in Table 2.1 and checking that each combination of variables results in the same value in each side of the equation. We shall use this method again in proving DeMorgan's theorem in Section 2.3.4.

Table 2.1 *Proving basic identities using a truth table*

A	$\bar{A}$	$A \cdot 1$	$A \cdot 0$	$A \cdot A$	$A + 1$	$A + 0$	$A + A$	$A \cdot \bar{A}$	$A + \bar{A}$	$\bar{\bar{A}}$
0	1	0	0	0	1	0	0	0	1	1
1	0	1	0	1	1	1	1	0	1	0

The basic identities can be extended to more variables, i.e.:

$$A \cdot B \cdot 1 = A \cdot B$$
$$A \cdot B \cdot 0 = 0$$
$$A \cdot B \cdot A = A \cdot B$$
$$A + B + 1 = 1$$
$$A + B + 0 = A + B$$
$$A + B + A = A + B$$
$$A \cdot B \cdot \bar{A} = 0$$
$$A + B + \bar{A} = 1$$

SELF-ASSESSMENT QUESTION 2.4

Prove the identity $A \cdot B \cdot 1 = A \cdot B$ using a truth table.

Applications

Let us here describe some "system-level" applications of the identities. (The basic identities are most often used to simplify Boolean expressions, which will be considered later.)

Unused inputs of an AND gate

An application of the basic identities is to determine what to do with unused inputs of gates. Suppose we have a three-input AND gate available, but simply require the two-input function, AB. A and B would be applied to two of the inputs. What do we do with the third input, say called C? Let us consider each of the following:

1. C connected to a permanent logic 1 (+5 V)
2. C connected to a permanent logic 0 (0 V)
3. C connected to A or B
4. Leave C disconnected.

Leaving C disconnected creates a condition on C which will depend upon the actual circuit design of the gate, which could vary with different circuit designs. The logical value on C could be a 0 or a 1. Also an unconnected input will be open to being affected by external influences such as electromagnetic radiation. So we shall disregard the fourth option as being impractical or at least undesirable.

Let us look at the effect of each of the other alternatives. If the unused input is connected to 1, we get the function $AB1 = AB$. If the unused input is connected to 0, we get the function $AB0 = 0$. Connecting the unused input to A, we get the function $ABA = AB$. To maintain the logic AND function between the two variables A and B, we have two choices:

1. C connected to a permanent logic 1 (+5 V)
2. C connected to A or B.

Which we choose will depend upon system factors. Connecting C to +5 V is the least disruptive to the system. Assuming that A (or B) is produced by a gate, connecting C to A (or B) increases the load that must be driven by this gate (i.e. three inputs instead of one). Gates can only be connected to a limited number of other gates, the number being called the *fan-out*. Figure 2.5 shows a situation of connecting C to A. Gate G_1 has to drive two inputs now.

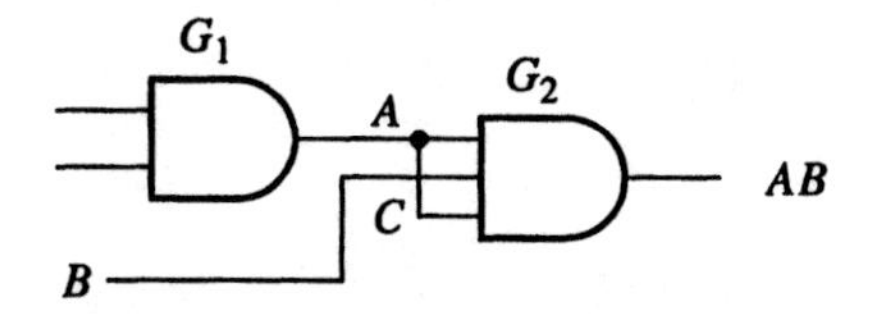

Figure 2.5 Handling unused inputs of AND gate, G_2

Unused inputs of an OR gate

Suppose we have a three-input OR gate available, but simply require the two-input function, $A + B$. A and B would be applied to two of the inputs. Let us consider each of the following:

1. C connected to a permanent logic 1 (+5 V)
2. C connected to a permanent logic 0 (0 V)
3. C connected to A or B.

If we connect the unused input to 1, we get the function $A + B + 1 = 1$. If we connect the unused input to 0, we get the function $A + B + 0 = A + B$. Connecting the unused input to A, we get the function $A + B + A = A + B$. To maintain the logic OR function between the two variables A and B, we have two choices:

1. C connected to a permanent logic 0 (0 V)
2. C connected to A or B.

Connecting C to 0 V is the simplest. Figure 2.6 shows a situation of connecting C to A. Again gate G_1 has to drive two inputs now.

Gate Faults

Another application of the basic identities is determining the effect of faults on gates. Knowing about potential faults in a circuit is very important for a logic designer. Faults (not design errors!) can occur during manufacture or subsequently. Faults can often be modelled by assuming that a permanent logic 0 or a permanent logic 1 appears on a signal line (the so-called *stuck-at fault model*).[2]

Suppose we have a two-input AND gate having the inputs A and B. What happens if a fault occurs? Suppose there are three possible gross fault conditions that could occur on the inputs, either:

1. A permanent 1
2. A permanent 0, or
3. A connection across the two inputs.

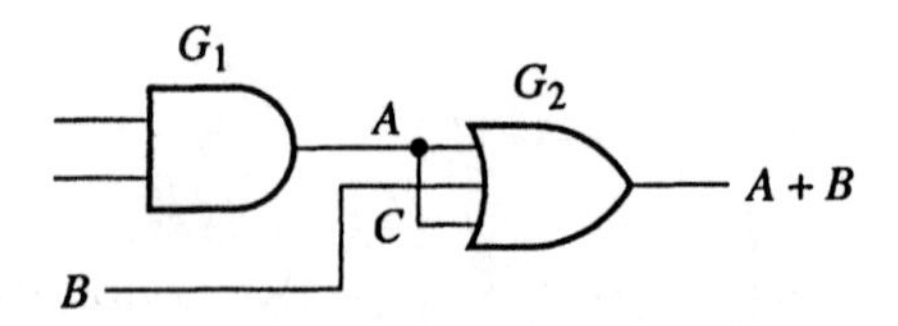

Figure 2.6 Handling unused inputs of OR gate, G_2

[2] Faults in logic circuits will be considered in detail in Chapter 7.

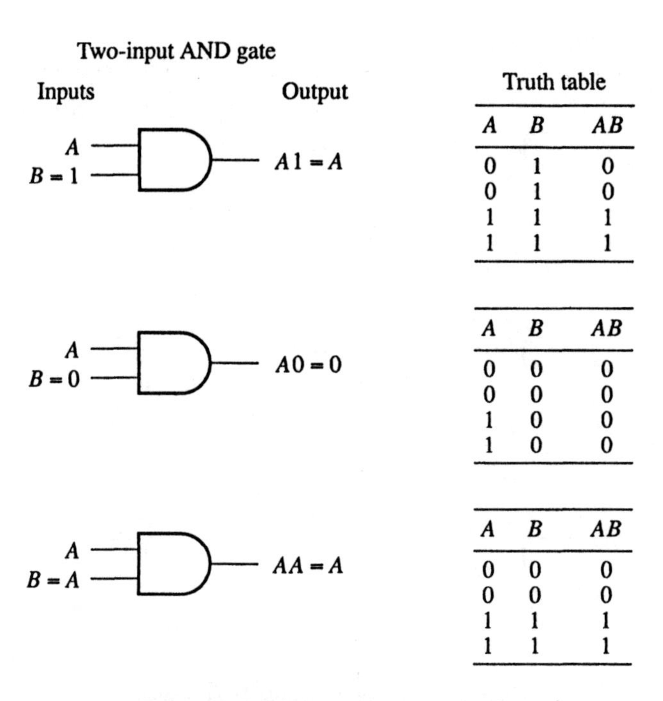

Truth table

A	B	AB
0	1	0
0	1	0
1	1	1
1	1	1

A	B	AB
0	0	0
0	0	0
1	0	0
1	0	0

A	B	AB
0	0	0
0	0	0
1	1	1
1	1	1

Figure 2.7 Faults on an input of an AND gate

The effect on the output can quite easily be found from the basic relationships. The effects are shown in Figure 2.7. A permanent 1 induces an output which will be the same as the other input. A permanent 0 induces a 0 output permanently, and a short across the inputs will produce the same effect as a permanent 1, i.e. an output which is same as the other input, A.

Suppose we have a two-input OR gate having the inputs A and B. We can find the effect of an input fault on the outputs again quite easily from the basic relationships. The effects are shown in Figure 2.8. A permanent 1 induces a permanent 1 output. A permanent 0 induces an output which will be the same as the other input. A short across the inputs will produce the same effect as a permanent 0, i.e. an output which is same as the other input. These results will be used to detect faults in logic circuits in Chapter 7.

2.3.2 Basic algebraic rules

In ordinary algebra, certain "rules" of algebra are assumed without thought. For example, $A + B = B + A$ and $A.B = B.A$ where + is arithmetic addition and . is arithmetic multiplication. It so happens that this rule also applies to both OR and AND operators in Boolean algebra, but only because the truth table of the Boolean operators are symmetrical, i.e. we can swap A and B in the table and still get a correct table for the function. So in Boolean algebra:

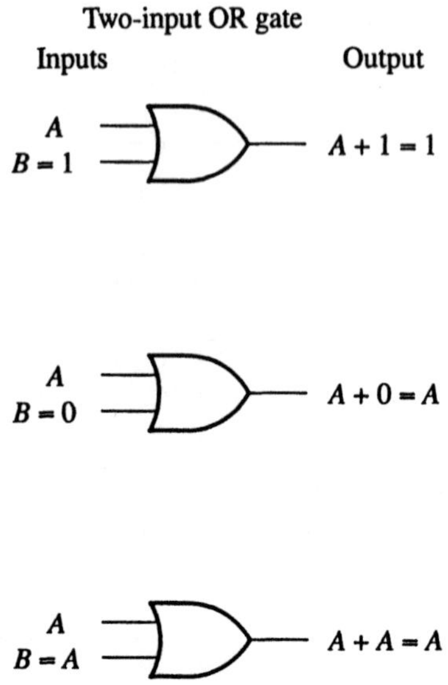

Two-input OR gate

Inputs — Output

A, $B = 1$ — $A + 1 = 1$

Truth table

A	B	$A + B$
0	1	1
0	1	1
1	1	1
1	1	1

A, $B = 0$ — $A + 0 = A$

A	B	$A + B$
0	0	0
0	0	0
1	0	1
1	0	1

A, $B = A$ — $A + A = A$

A	B	$A + B$
0	0	0
0	0	0
1	1	1
1	1	1

Figure 2.8 Faults on an input of an OR gate

$$A + B = B + A$$
$$AB = BA$$

This "rule" is formally called the *commutative law* (in either algebra).

The order that we compute an expression composed of variables connected by the same operator is not important in ordinary algebra. For example, in ordinary arithmetic, $ABC = (AB)C = A(BC)$ where the parentheses indicate which pair of variables are processed first. Similarly the order is not important in Boolean algebra, i.e. $ABC = (AB)C = A(BC)$. This rule is formally called in the *associative law.* It can be stated as:

$$ABC = (AB)C = A(BC)$$
$$A + B + C = (A + B) + C = A + (B + C)$$

In ordinary algebra, we can use parentheses to enforce a particular order of computation, which may be necessary when the operators are different. Order can be enforced with parentheses in Boolean algebra as we can see above. Parentheses are also used in ordinary algebra to show that multiplication applies to a group. For example $A(B + C)$ means A is multiplied by the summation of B and C. We can expand out the expression to $AB + AC$, i.e. the multiplication distributed across both B and C. This is also true in Boolean to both AND and OR operators and is called the *distributive law.* In Boolean algebra, the distributive law states that:

$$A(B + C) = AB + AC$$
$$A + (BC) = (A + B)(A + C)$$

Notice how the OR operation "distributes" through both B and C terms. The distributive law does not apply to arithmetic addition, i.e. $A + (BC) \neq (A + B)(A + C)$ in ordinary algebra.

Applications of basic laws to logic design

The laws and relationships so far give some tools for designing a logic expression and creating alternatives. For example, the associative law shows us that we can factorize an expression to obtain different implementations such as is shown in Figure 2.9 for $f = ABCD$. This may be important if the gates available have a particular number of inputs. (The number of inputs of a gate is usually between 2 and 8.) In Figure 2.9(a), a single 4-input AND gate is used; in Figure 2.9(b) and (c) "two-level" arrangements are used. In two-level circuits, the signals may pass through two gates from the inputs to the outputs. Two-level circuits are particularly common for implementing expressions, see Chapter 3.

The distributive law shows us that the function such as $f = AB + AC$ has at least two implementations as shown in Figure 2.10. Obviously one, Figure 2.10(b), requires less gates. The concept of creating a circuit with the least number of gates is of course very important and comes under the topic of *logic minimization*, which we shall examine in detail in Chapter 3.

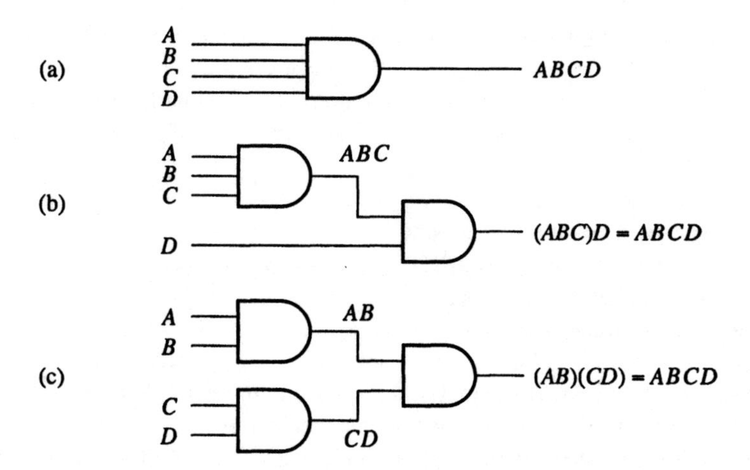

Figure 2.9 Using the associative law for implementing the function f = ABCD

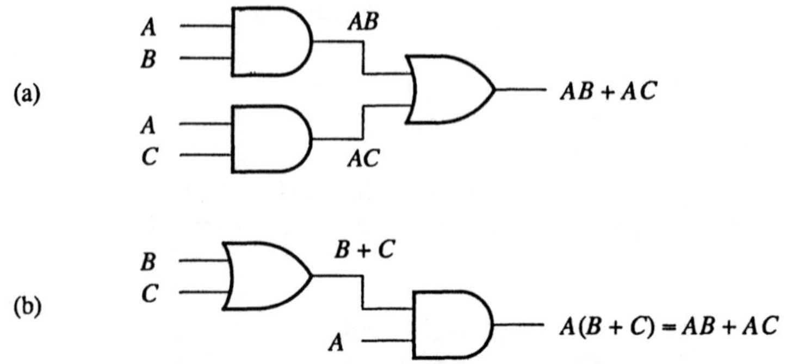

Figure 2.10 Using the distributive law for implementing the function $f = AB + AC$

SELF-ASSESSMENT QUESTION 2.5

Draw two alternative logic implementations for the function $f = A\overline{B}C + \overline{B}D$.

There are some further basic Boolean laws which can help in rearrangement and minimization. Let us look at those now.

2.3.3 *Duality*

Dual functions

If we change all the 0's to 1's and all the 1's to 0's in the AND truth table, we get the OR truth table. Hence if we have a circuit which implements the AND function using positive logic representation, where logic 0 is represented by 0 V and a logic 1 is represented by +5 V, the same circuit accepting the same voltage levels would perform the OR function using the negative logic representation where a 0 is represented by +5 V and a logic 1 is represented by 0 V. No changes to the circuit would be needed. Similarly an OR gate with positive logic representation would change to an AND gate with negative logic representation. A NOT gate is a NOT gate in both representations.

Suppose we have a logic circuit which performs the function $f = A + B\overline{C}$ with positive logic representation. The same circuit performs the function $f = A(B + \overline{C})$ (i.e. with the operators swapped, AND to OR, OR to AND) with negative logic representation. This phenomenon is shown in Figure 2.11. Further, if we have a complex logic circuit performing some function $f(A, B, C \ldots \cdot, +, 0, 1)$ with positive logic representation, i.e. a function having the variables $A, B, C \ldots$, AND and OR operations, and permanent 0's and permanent 1's, the same circuit would perform the function $f(A, B, C \ldots, +, \cdot, 1, 0)$, with negative logic representation, i.e. a function with all +'s changed to ·'s, all ·'s changed to +'s, all 0's changed to 1's and all 1's changed to 0's. The circuit need not be designed just with AND, OR and NOT gates; more complex circuit arrangements could be used, but the effect is the same because the circuit can

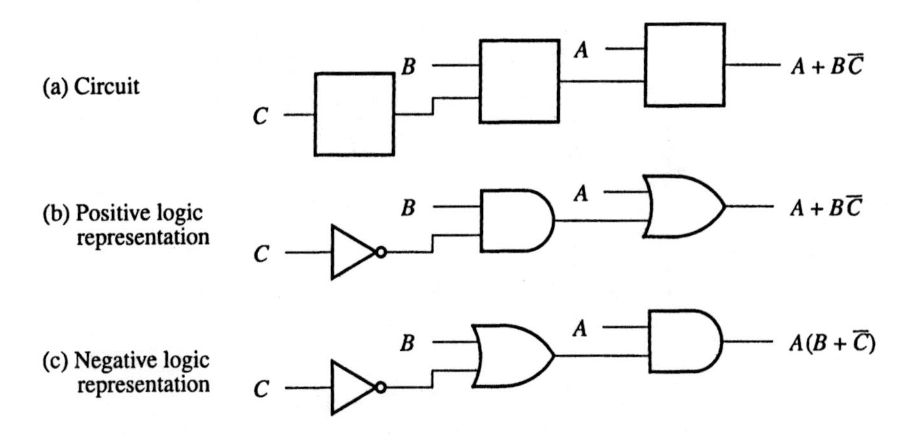

Figure 2.11 Duality of circuits

always be reduced to AND, OR and NOT circuits and we are simply applying voltages to the inputs and producing voltages at the output of the circuit.

The function produced with negative logic representation is the *dual function* to that produced by positive logic representation, i.e. the dual of the function f:

$$f = f(A, B, C \ldots \cdot, +, 0, 1)$$

is:

$$f^{\mathrm{d}} = f(A, B, C \ldots, +, \cdot, 1, 0)$$

SELF-ASSESSMENT QUESTION 2.6

Derive the dual expression to $f = A + B(C + AD)$.

Dual equalities — Principle of Duality

Suppose we have proved an equality:

$$f(A, B, C \ldots \cdot, +, 0, 1) = g(A, B, C \ldots \cdot, +, 0, 1)$$

It then follows that a dual equality:

$$f^{\mathrm{d}}(A, B, C \ldots \cdot, +, 1, 0) = g^{\mathrm{d}}(A, B, C \ldots \cdot, +, 1, 0)$$

is also true. This is known as the *Principle of Duality.* It can be proved by considering positive and negative logic representation. Given two logic circuits, L_1 and L_2, if they produce equivalent functions $f()$ and $g()$ with positive logic representation, they will produce the equivalent functions $f^{\mathrm{d}}()$ and $g^{\mathrm{d}}()$ with negative logic representation, as illustrated in Figure 2.12.

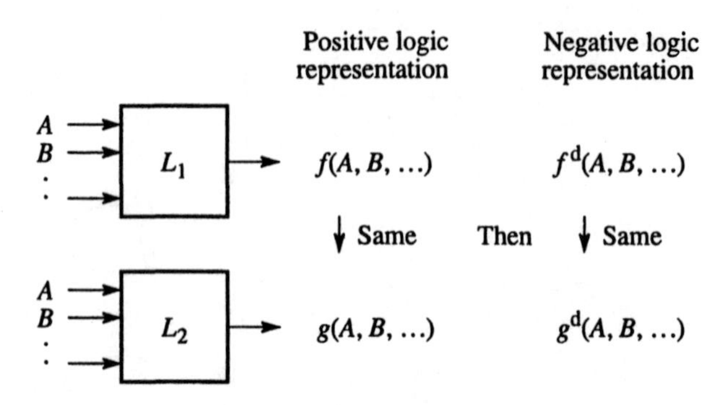

Figure 2.12 Dual equalities

To obtain the dual equality, we simply change all +'s to ·'s, all ·'s to +'s all 0's to 1's and all 1's to 0's. The basic identities can be grouped into duals.

EXAMPLES

(a) Suppose we have proved an equality such as:

$$A + 0 = A$$

The dual expression of $A + 0$ must also equal the dual expression of A according to the Principle of Duality. The dual of $A + 0$ is $A1$ and the dual of A is A. Therefore the dual equality:

$$A1 = A$$

is true without proof.

(b) Suppose we have proved that:

$$A + A\overline{B} = A$$

which is easy because $A + A\overline{B} = A(1 + \overline{B}) = A$.[3] We can immediately write the dual equality:

$$A(A + \overline{B}) = A$$

by changing all the operators. Interesting in this example, if we expand the left side we get the first expression, i.e. $A(A + \overline{B}) = AA + A\overline{B} = A + A\overline{B}$, which all equal A.

[3] Notice here algebraic factorization is being used, and a basic identity.

SELF-ASSESSMENT QUESTION 2.7

Derive the dual equality to $AB + \bar{A}C + BC = AB + AC$ (the Consensus theorem, see Chapter 3).

2.3.4 DeMorgan's theorem

The OR function $A + B$ can only be a 0 in one instance, when both A and B are a 0. In all other instances, $A + B$ is a 1. Similarly, the AND function AB can only be a 1 in one instance, when both A and B are a 1. In all other instances, AB is a 0. Clearly there must be a relationship between the AND function and the OR function. The function $\bar{A} + \bar{B}$ is a 0 in one instance, when $A = 1$ and $B = 1$. Similarly the function $\overline{AB}$ is also a 0 in one instance, when $A = 1$ and $B = 1$. Therefore:

$$\bar{A} + \bar{B} = \overline{AB}$$

and the dual:

$$\bar{A}\bar{B} = \overline{A + B}$$

are true. These relationships are know as *DeMorgan's theorem* (given here for two variables). DeMorgan's theorem can be proved by listing each function in a truth table and confirming that both are the same for each combination of A and B as shown in Table 2.2. As we have seen, the concept of using a truth table to prove an identity is reasonable and powerful if there are only a few variables.

Table 2.2 *Proving DeMorgan's theorem for two variables*

A	B	$A+B$	AB	$\bar{A}+\bar{B}$	$\overline{AB}$
0	0	0	0	1	1
0	1	1	0	1	1
1	0	1	0	1	1
1	1	1	1	0	0

DeMorgan's theorem can be extended to more variables. For three variables, we have:

$$\bar{A} + \bar{B} + \bar{C} = \overline{ABC}$$
$$\bar{A}\bar{B}\bar{C} = \overline{A + B + C}$$

and more generally:

$$\bar{A} + \bar{B} + \bar{C} \ldots = \overline{ABC} \ldots$$
$$\bar{A}\bar{B}\bar{C} \ldots = \overline{A + B + C} \ldots$$

A proof of DeMorgan's theorem for any number of variables can start with the proof of DeMorgan's theorem for two variables using the truth table approach as shown in Table 2.2 which lists all combinations of variables and their corresponding results. Given that this equality is proved, it can be used with different named variables without further proof.

By substituting BC for B in:

$$\bar{A} + \bar{B} = \overline{AB}$$

where C is a third variable, we get:

$$\bar{A} + (\overline{BC}) = \overline{A(BC)}$$

We already know that $\overline{BC} = \bar{B} + \bar{C}$ from DeMorgan's theorem for two variables. Substituting, we get DeMorgan's theorem for three variables:

$$\bar{A} + \bar{B} + \bar{C} = \overline{ABC}$$

Clearly this process can be repeated for four and more variables. Hence DeMorgan's theorem must be true for any number of variables ("proof by induction"). A similar approach can be used to prove the dual equality, if we do not want to rely on the Principle of Duality.

SELF-ASSESSMENT QUESTION 2.8

Prove the dual form of DeMorgan's theorem for three variables, i.e.: $\bar{A}\bar{B}\bar{C} = \overline{A + B + C}$.

DeMorgan's theorem can be generalized even further to include AND and OR on both sides of the equality. This general form of DeMorgan's theorem can be written as:

$$\bar{f}(A, B, C \ldots +, \cdot) = f(\bar{A}, \bar{B}, \bar{C} \ldots \cdot, +)$$

or:

$$f(A, B, C \ldots +, \cdot) = \bar{f}(\bar{A}, \bar{B}, \bar{C} \ldots \cdot, +)$$

which means that a function, f, consisting of variables and AND and OR operators, is equivalent to the inverse function consisting of the variables inverted (complemented) and the operators reversed (AND to OR, OR to AND).

Be careful not to confuse DeMorgan's theorem with the Principle of Duality. The

Principle of Duality allows us to write an equality as a dual of another equality. None of the variables is complemented in the process of creating the dual. DeMorgan's theorem states a specific equality and a dual equality which may be used to simplify expressions.

SELF-ASSESSMENT QUESTION 2.9

Apply the general form of DeMorgan's theorem to $f = A + B + \bar{C}D$.

2.4 Universal gates

2.4.1 NAND gate

DeMorgan's theorem provides a relationship between AND and OR. Hence, of the three "basic" operations, AND, OR and NOT, only two are strictly necessary (AND and NOT or OR and NOT) because the third can be replaced by a function of the other two. From DeMorgan's theorem we have:

$$A + B = \overline{\bar{A}\bar{B}}$$

Hence $A + B$ can be replaced by $\overline{\bar{A}\bar{B}}$. We now have two operators, AND and NOT. We can actually combine AND and NOT to form a "universal" operation, NOT–AND or NAND, which can be used to implement AND, OR and NOT and hence any Boolean expression. The NAND operation with two variables, A and B, is:

$$A \text{ "NAND" } B = \overline{AB}$$

The "distinctive shape" symbols for two-input and three-input NAND gates and their corresponding truth tables are shown in Figure 2.13.

It is a simple matter to show that all three basic operations, AND, OR and NOT, can be obtained from NAND. The NOT operation can be obtained from NAND by replacing B with A in the above to obtain $\overline{AA}$, i.e.:

$$\overline{AA} = \bar{A}$$

The OR operation can be obtained by substituting the above, i.e.:

$$A + B = \overline{(\overline{AA})(\overline{BB})}$$

Similarly the AND operation can be obtained:

$$AB = (\overline{\overline{AB}})$$

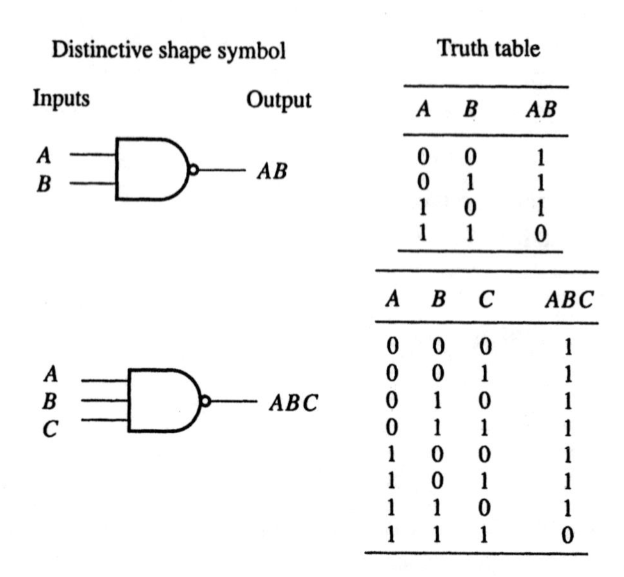

Truth table

A	B	AB
0	0	1
0	1	1
1	0	1
1	1	0

A	B	C	ABC
0	0	0	1
0	0	1	1
0	1	0	1
0	1	1	1
1	0	0	1
1	0	1	1
1	1	0	1
1	1	1	0

Figure 2.13 NAND gate symbol and truth table

The results of these substitutions are shown in Figure 2.14. It follows that the NAND function, $\overline{AB}$, could be used to implement any AND/OR/NOT function, i.e. any Boolean function at all. In this sense, the NAND function $\overline{AB}$ is "universal", and a gate that performs this function would also be universal in that only this type of gate would be needed to implement any logic circuit. Only one type would be needed for inventory and for maintenance. The concept of a universal gate was very attractive when logic systems had to be constructed from packaged gates. It also turned out that the electronic circuits for creating a NAND function is slightly simpler and faster than the circuits for AND and OR. Finally, a direct substitution as in Figure 2.14 would not necessarily be done; the overall circuit would be designed with NAND gates in mind.

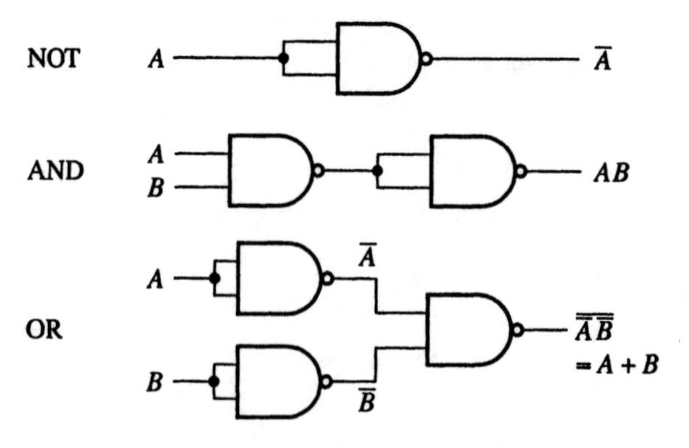

Figure 2.14 NOT, AND and OR from NAND gates

2.4.2 NOR gate

Instead of using the NAND gate as the universal gate, we could use the function $\overline{A+B}$ as the universal function. This function is the NOR function (NOT–OR) and the gate to implement this function, the NOR gate. From DeMorgan's theorem we have:

$$AB = \overline{\overline{A}+\overline{B}}$$

Hence AB can be replaced with $\overline{\overline{A}+\overline{B}}$. The NOT operator can be obtained by replacing B with A in $\overline{A+B}$ to get $\overline{A+A} = \overline{A}$. Hence again we have a universal operator. The "distinctive shape" symbols for two-input and three-input NOR gates and their corresponding truth tables are shown in Figure 2.15. Generating AND and OR from NOR is shown in Figure 2.16. A special logic symbol is not commonly used in engineering to indicate either NAND or NOR, though mathematical symbols do exist (↑ and ↓ for NAND and NOR respectively).

There is a duality between NAND and NOR. A gate that performs a NAND operation in positive logic representation performs the NOR operation in negative logic representation (and vice versa). This can be proved by examining the truth tables of NAND and NOR. If all the 0's were changed to 1's and all the 1's changed to 0's, we get the other operation, as shown in Table 2.3.

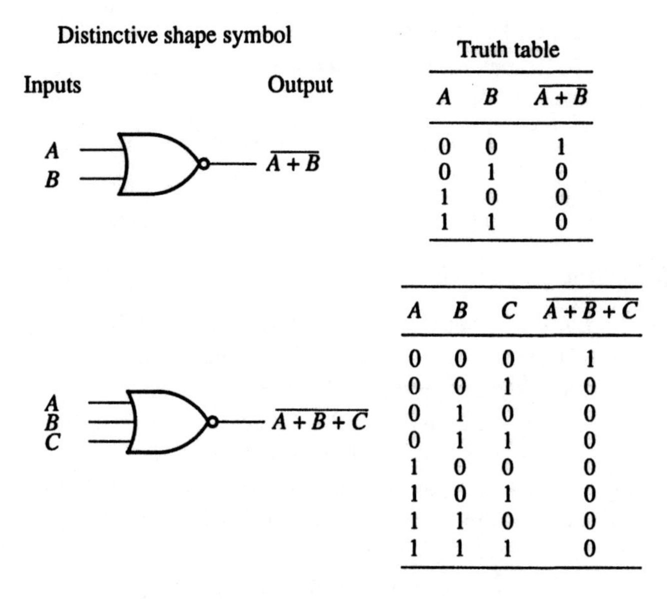

A	B	$\overline{A+B}$
0	0	1
0	1	0
1	0	0
1	1	0

A	B	C	$\overline{A+B+C}$
0	0	0	1
0	0	1	0
0	1	0	0
0	1	1	0
1	0	0	0
1	0	1	0
1	1	0	0
1	1	1	0

Figure 2.15 NOR gate symbol and truth table

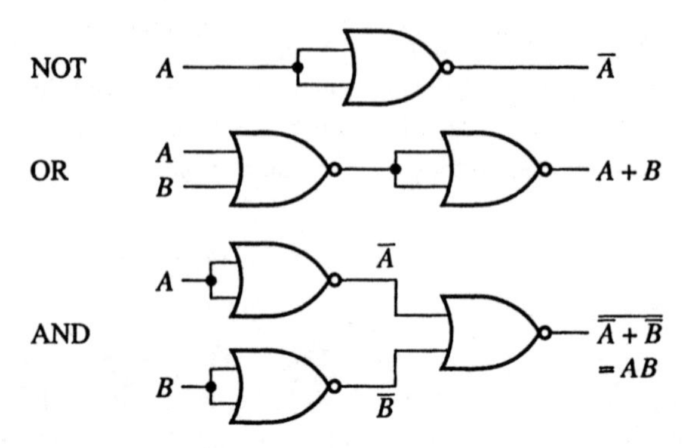

Figure 2.16 NOT, OR and AND from NOR gates

Table 2.3 *Duality between NAND and NOR operators*

NAND			*NOR*		
Inputs		*Output*	*Inputs*		*Output*
0	0	1	1	1	0
0	1	1	1	0	0
1	0	1	0	1	0
1	1	0	0	0	1

SELF-ASSESSMENT QUESTION 2.10

Show how the NOR function can be obtained from NAND gates, and how a NAND function can be obtained from NOR gates.

2.5 Other gates

2.5.1 Exclusive-OR gate

The OR operation, $A + B$, when producing a 1 *includes* when both A and B are a 1. More accurately this OR operator should be called the *inclusive*-OR operator. The "OR" operation which does not produce a 1 when A and B are both a 1, i.e. produces a 1 if $A = 1$ or $B = 1$ but not both A and B a 1, is called the *exclusive*-OR operator. The exclusive-OR gate symbol and truth table are shown in Figure 2.17. Note the similarity of the symbol with the normal (inclusive) OR gate. The exclusive-OR operation is indicated with the ⊕ symbol. In terms of two operands, A and B, the exclusive-OR operation is:

$$A \oplus B = A\overline{B} + \overline{A}B$$

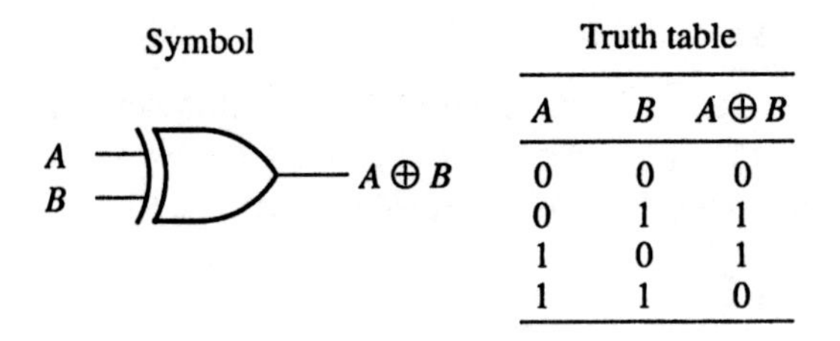

A	B	$A \oplus B$
0	0	0
0	1	1
1	0	1
1	1	0

Figure 2.17 Exclusive-OR gate and truth table

i.e. is a 1 when either A is a 1 and B is a 0 ($A\overline{B}$), or A is a 0 and B is a 1 ($\overline{A}B$). These two conditions can be identified in the truth table where the result is a 1. Any function can be obtained from a truth table by looking for the 1's in the output. The corresponding combination variables become one of the terms of the function.

The (inclusive) OR gate can have more than two inputs. The exclusive-OR gate is limited to having two operands. (How would one define the function with more than two operands? Would two 1's result in a 1, or three 1's ...?)

SELF-ASSESSMENT QUESTION 2.11

Suppose a 3-input "exclusive-OR" gate were invented to produce a 1 if $A = 1$ or $B = 1$ or $C = 1$ but not any combination of A, B or C a 1 together. What would the function be?

The exclusive-OR operation is not regarded as a fundamental operation as the operation can be broken down into AND, OR and NOT operations. However, there are basic identities for the exclusive-OR operation just as there are for AND, OR and NOT:

$$A \oplus 0 = A$$
$$A \oplus 1 = \overline{A}$$
$$A \oplus A = 0$$
$$A \oplus \overline{A} = 1$$
$$A \oplus \overline{B} = \overline{A} \oplus B = \overline{(A \oplus B)}$$

The exclusive-OR operation also obeys the commutative, associative and distributive laws:

Commutative law: $A \oplus B = B \oplus A$
Associative law: $(A \oplus B) \oplus C = A \oplus (B \oplus C) = A \oplus B \oplus C$
Distributive law: $A(B \oplus C) = AB \oplus AC$

Proof of these can be done using truth tables, or by expanding out the exclusive-OR operation into basic operations.

2.5.2 Exclusive-NOR gate

The inverse operation of the exclusive-OR is called the exclusive-NOR operation. The exclusive-NOR gate symbol and truth table are shown in Figure 2.18. In terms of two operands, A and B, the exclusive-NOR operation is:

$$(\overline{A \oplus B}) = AB + \bar{A}\bar{B}$$

i.e. is a 1 when $A = 1$ and $B = 1$ (AB) or $A = 0$ and $B = 0$ ($\bar{A}\bar{B}$). This operation is useful because it compares two binary digits and returns a 1 only when both digits are the same (i.e. either both a 1 or both a 0), and hence can implement a comparator or equivalence function. The exclusive-OR operation results in a 1 when the digits are different (a not-equivalence function). The exclusive-NOR operation also obeys the commutative and associative laws. It is left as an exercise to determine whether it obeys the distributive law. As with NAND and NOR, a specific symbol is not used to indicate the exclusive-NOR operation.

2.5.3 Commutative functions

So far, six functions of two variables have been identified, namely AND, OR, NAND, NOR, exclusive-OR and exclusive-NOR. In fact, there are sixteen different functions of two Boolean variables, including permanent 1 and permanent 0, as listed in Table 2.4. Of these functions, F_1, F_6, F_7, F_8, F_9 and F_{14} are commutative functions, i.e. the variables can be interchanged without affecting the results, ignoring F_0 (permanent 0) and F_{15} (permanent 1). F_1 is the AND function, F_6 the exclusive-OR function, F_7 the OR function, F_8 the NOR function, F_9 the exclusive-NOR function, F_{14} the NAND function. The remaining functions are not commutative. Hence we see that AND, OR, exclusive-OR, exclusive-NOR, NAND and NOR form the complete set of commutative functions. Gates are available for these commutative functions and are not usually available for the other non-commutative functions unless as part of a specialized component. One advantage of limiting gates to commutative functions is that the inputs can be used in any order, and interchanged at will.

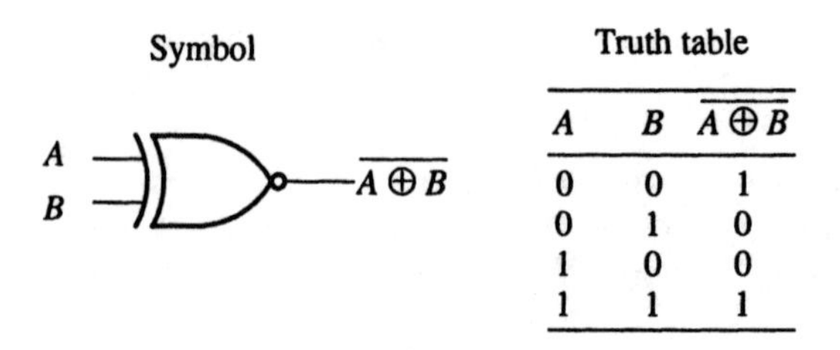

A	B	$\overline{A \oplus B}$
0	0	1
0	1	0
1	0	0
1	1	1

Figure 2.18 Exclusive-NOR gate symbol and truth table

Table 2.4 *Functions of two variables*

AB	F_0	F_1	F_2	F_3	F_4	F_5	F_6	F_7	F_8	F_9	F_{10}	F_{11}	F_{12}	F_{13}	F_{14}	F_{15}
0 0	0	0	0	0	0	0	0	0	1	1	1	1	1	1	1	1
0 1	0	0	0	0	1	1	1	1	0	0	0	0	1	1	1	1
1 0	0	0	1	1	0	0	1	1	0	0	1	1	0	0	1	1
1 1	0	1	0	1	0	1	0	1	0	1	0	1	0	1	0	1

SELF-ASSESSMENT QUESTION 2.12

What feature of the entries in Table 2.4 demonstrates that the functions F_2, F_3, F_4, F_5, F_{10}, F_{11}, F_{12}, and F_{13} are not commutative?

2.6 Gate design

In this section, we shall outline the internal features of gates, first briefly reviewing the now very old but still used TTL family, and then the MOS gate in somewhat more detail. This section could be omitted without loss of continuity.

2.6.1 TTL combinational logic circuits

Gates are constructed using semiconductor integrated circuit technology. Single gates are not manufactured in one integrated circuit package, as the technology offers the capability to manufacture many gates in one package. The first highly successful type of integrated circuit family of gates was the TTL (transistor–transistor logic) family introduced in the 1960s and still used for small digital systems and for interfaces to larger integrated circuit packages. In many cases, the original TTL technology has been replaced with CMOS technology (see Section 2.7), but the functionality of the packages remains the same.

TTL devices are usually manufactured with a few gates in one package, typically between 2 and 100. The term *small-scale integration* (SSI) is used when the number of gates is less than about 12. The term *medium-scale integration* is used when the number of gates is between about 12 and 100. The term *large-scale integration* is (LSI) used for integrated circuit logic devices when the number of gates is between about 100 and 1000, the term *very large-scale integration* (VLSI) being used when there are thousands of gates in the packaged device. At this level, the number of transistors rather than the number of gates may be quoted. A single gate may require about 2–6 transistors depending upon the technology and the gate function. TTL requires four transistors for a basic NAND gate.

TTL is typically packaged in "dual-in-line" packages, the smallest measuring about 0.75 of an inch long by 0.25 of an inch wide and having seven pins along two

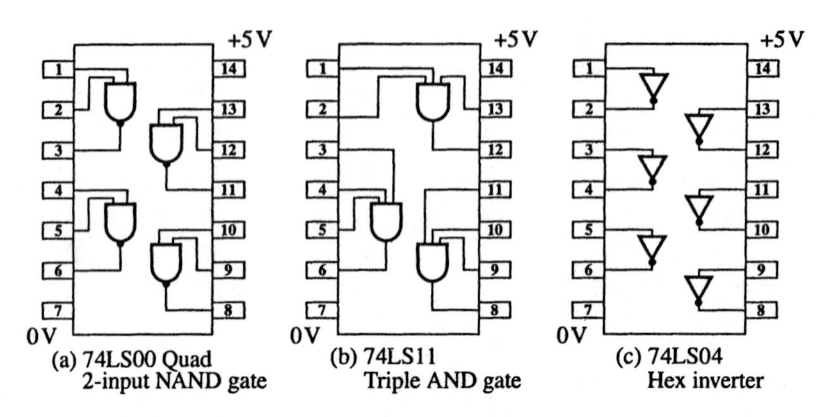

Figure 2.19 Dual-in-line TTL package layouts

opposite edges (i.e. dual pins in line) spaced at 0.1 inch intervals. The package is about 0.2 of an inch high. This package can be used for a few gates, examples of which are shown in Figure 2.19. These examples are from the TTL 74LSXX series. Longer dual in-line packages exist having more pins along two edges. A list of popular devices is given in Table 2.5. There is a large number of packaged SSI/MSI devices in the TTL family in addition to those given in Table 2.5. Some devices perform common functions such as decoding; such devices are described in Chapter 3.

In TTL, a logic 1 is represented by a voltage of +3.4 V nominally, though it can be generated between +2.4V and +5V. We often associated +5 V with a logic 1. The supply voltage for TTL is also +5 V. A logic 0 is represented by the voltage of +0.2 V nominally, though it can be generated between 0 V and +0.4 V. We often associate 0 V with a logic 0. There may be slight variations in nominal voltages between logic device types. Voltages as low as +2 V will be recognized as logic 1 and voltages as high as +0.8 V will be recognized as logic 0, to allow extraneous electrical "noise" in the system. The overlap between the recognized voltage ranges and the generated voltage ranges is shown in Figure 2.20.

Noise is the term used to describe unwanted electrical signals occurring on wires in a system. It comes about from the normal switching operation of logic gates, which can generate interference in neighbouring devices and on lines either by electromagnetic radiation or via associated power supply variations. Less commonly, the source of the noise can be external to the system. Logic devices must be designed to accept a certain amount of electrical noise in the system and continue to operate correctly. The *noise margin* is the level of voltage present as electrical noise that can be tolerated in the system. It is given in terms of the allowable noise voltage that can be added to or subtracted from a generated logic signal with the logic signal still recognized at the input of gate as a logic level. In TTL, the noise margins at both a logic 0 and a logic 1 is +0.4V. The noise margins given here are applicable to continuous low frequency noise. The behaviour of logic devices under the influence

Table 2.5 *Some early TTL devices*

Device number	*Gate type*	*Number of inputs*	*Number of gates*
74LS04	NOT	1-input	6
74LS08	AND	2-input	4
74LS11	AND	3-input	3
74LS21	AND	4-input	2
74LS32	OR	2-input	4
74LS00	NAND	2-input	4
74LS10	NAND	3-input	3
74LS20	NAND	4-input	2
74LS30	NAND	8-input	1
74LS133	NAND	13-input	1
74LS02	NOR	2-input	4
74LS27	NOR	3-input	3
74LS260	NOR	5-input	2
74LS86	Exclusive-OR	2-input	4
74LS266	Exclusive-NOR	2-input	4
74LS51	AND-OR-INVERT	2-wide 2/3-input	2
74LS54	AND-OR-INVERT	4-wide 2/3-input	1

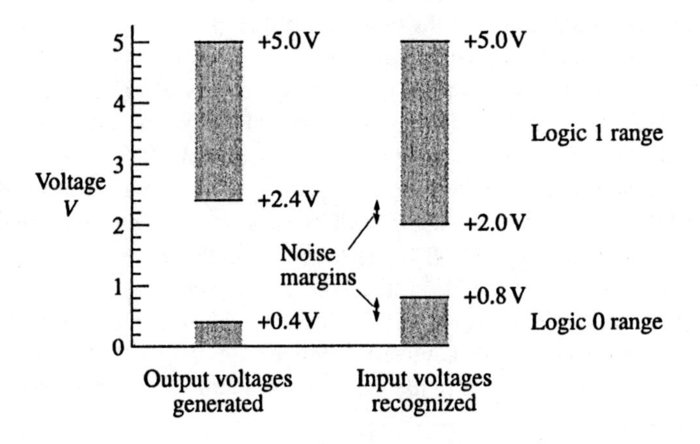

Figure 2.20 TTL voltages

of non-continuous or very high frequency noise may be considerably different to that of low frequency noise. The behavior of a logic device under high frequency noise conditions can be described by the *a.c. noise margins* as opposed to the d.c. noise margins.

2.6.2 Metal oxide silicon (MOS) gates

In this section, we shall outline the internal design of MOS (metal oxide silicon) gates. MOS gates have replaced TTL gates in many instances and are used in very large scale integrated circuits because of their much lower power consumption.

MOS transistor switches

A logic gate accepts one of two voltages and generates one of two voltages. Hence we sometimes refer to the voltages as a logic high voltage and a logic low voltage, say H and L. Each voltage could be generated by the mechanical switch arrangement shown in Figure 2.21. One switch connects the H voltage to the output and one switch connects the L voltage to the output. Only one switch is closed at any instant. To create an H voltage, the upper switch is closed and the lower switch is open. To create an L voltage, the lower switch is closed and the upper switch is open. Of course, mechanical switches are not used in our systems. However, the mechanical switch configuration of Figure 2.21 is the basis of most electronic gates.

Let us review the operation of the MOS transistor as it relates to a switch.[4] The MOS transistor has three "terminals", the *source* terminal, the *drain* terminal, and the *gate* terminal. The source and drain terminals correspond to the two terminals of a mechanical switch. Just as a mechanical switch has two states, one in which no electrical conduction occurs between its terminals and one in which electrical conduction occurs between its terminals, an MOS transistor can behave in a similar way. In a mechanical switch, to change the state of the switch, the switch arm is moved. In an MOS transistor, to change the state of the transistor, a different voltage is applied to the gate terminal of the transistor. When the transistor is switched "on", the electrical conduction occurs between the source and drain terminals. When the transistor is switched "off", no electrical conduction occurs between the source and drain terminals. The voltage on the gate to control the on and off states is actually defined as the voltage across the gate and source terminals, V_{gs}.

There are two types of MOS transistor, the *n*-type MOS transistor (NMOS transistor), and the *p*-type MOS transistor (PMOS transistor). Let us start with the *n*-type MOS transistor. The *n*-type MOS transistor requires the drain to be more positive than the source for correct operation. The voltage between the gate and source to turn the transistor "off" (create no conduction between the source and drain) is 0 V. To turn the transistor "on" (create conduction between the source and drain), a positive voltage has to be applied between the gate and the source. The voltage has to exceed

[4] More details and other types of logic circuits such as TTL (transistor–transistor logic) circuits can be found in textbooks given at the end of the chapter.

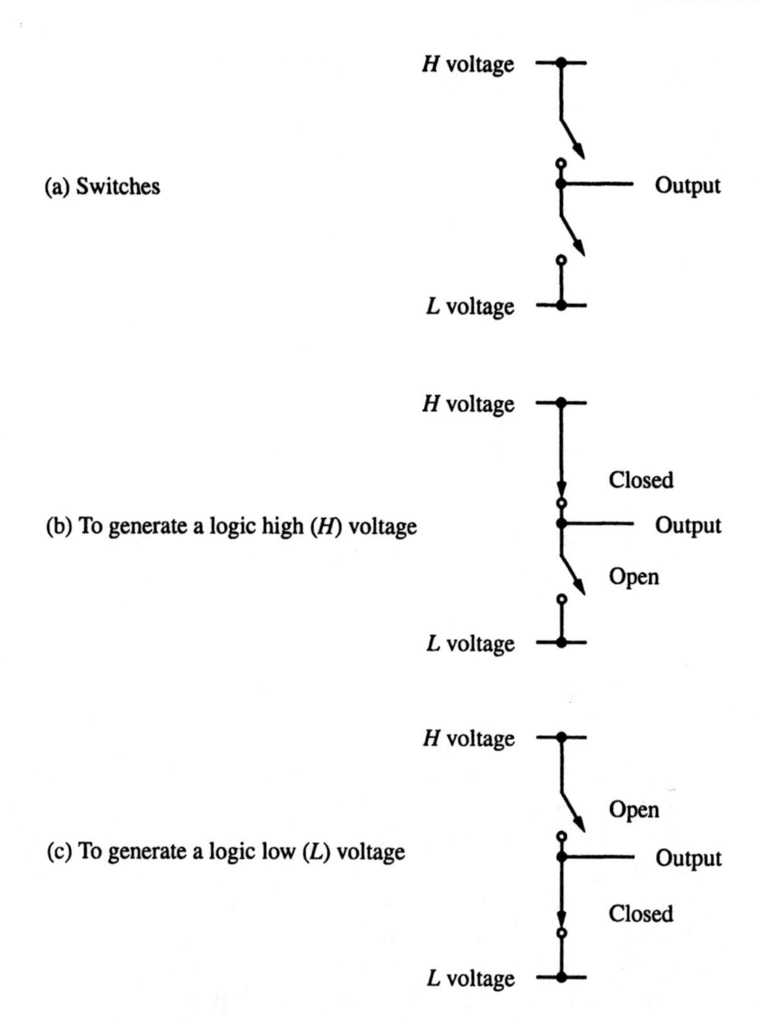

Figure 2.21 Complementary switches to produce logic high and low voltages

some *threshold voltage*. The construction of the transistor is chosen to obtain a specific threshold voltage. Typically the threshold voltage is +1.5V for circuits designed to operate with a +5V supply. Figure 2.22 shows the symbol of the *n*-type MOS transistor[5] and the applied voltages to achieve the two states, on and off. Here we are using +5V to turn the device on.

The *p*-type MOS transistor requires the drain to be more negative than the source for correct operation. The voltage between the gate and source to turn the transistor "off" (create no conduction between the source and drain) is again 0V. To turn the

[5] The symbol here is for a so-called enhancement mode *n*-type MOS transistor. There is another type of MOS transistor called the depletion mode MOS transistor which is not often used for switching circuits.

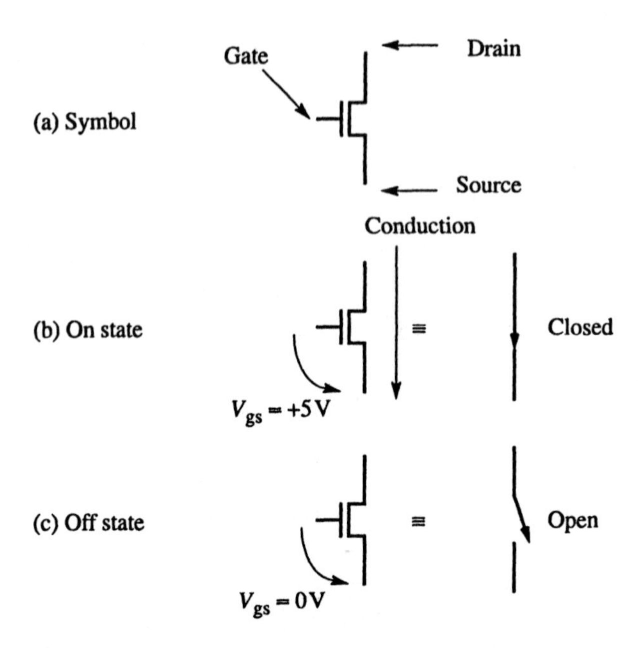

Figure 2.22 n-Type MOS transistor as a switch

transistor "on" (create conduction between the source and drain), a negative voltage has to be applied between the gate and source beyond the threshold voltage for the device typically −1.5 V. The symbol and applied voltages for the *p*-type MOS transistor are shown in Figure 2.23. In this example, −5 V will turn the device on. Notice that a bubble is used in the *p*-type transistor symbol. As we have seen, bubbles are also used in logic gate symbols to show inversion.

The switch configuration in Figure 2.21 can be created by using one *n*-type MOS transistor and one *p*-type MOS transistor as shown in Figure 2.24.[6] When the input is at 0 V, the gate–source voltage of the *n*-type transistor is also at 0 V, turning this transistor off. The gate–source voltage of the *p*-type MOS transistor is at −5 V. The source to gate voltage is +5 V, turning this transistor on. This creates the equivalent of the switch positions in Figure 2.21(b), and an output of +5 V. When the input is at +5 V, the gate–source voltage of the *n*-type MOS transistor is also at +5 V, turning this transistor on. The gate–source voltage of the *p*-type MOS transistor is at 0 V turning this transistor off. (The gate is at the same voltage as the source.) This creates the equivalent of the switch positions in Figure 2.21(c), and an output of 0 V.

The circuit configuration in Figure 2.24 is known as *complementary MOS* because the two transistors complement each other; when one is off the other is on and vice versa. Normally, we consider the inputs and outputs of logic circuits in terms of logic

[6] A specific supply voltage of +5 V is used here to help understand the operation of the circuit. The supply voltage can be given the notation, V_{DD}, and can be lower than +5 V. 0 V can be given the notation V_{SS}.

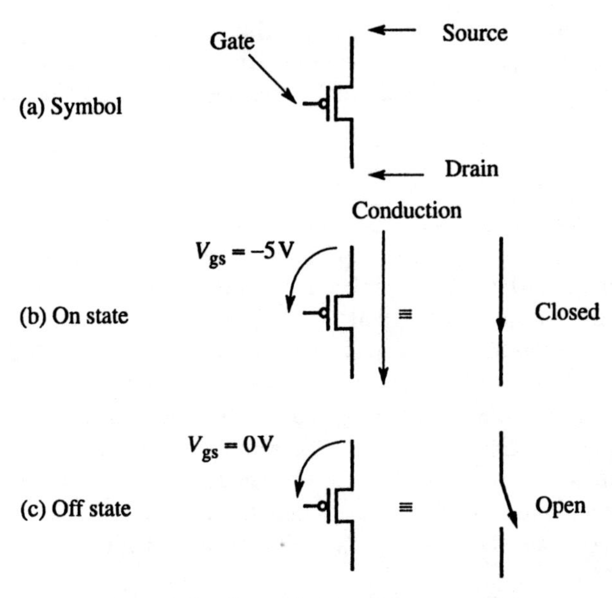

Figure 2.23 p-Type MOS transistor as a switch

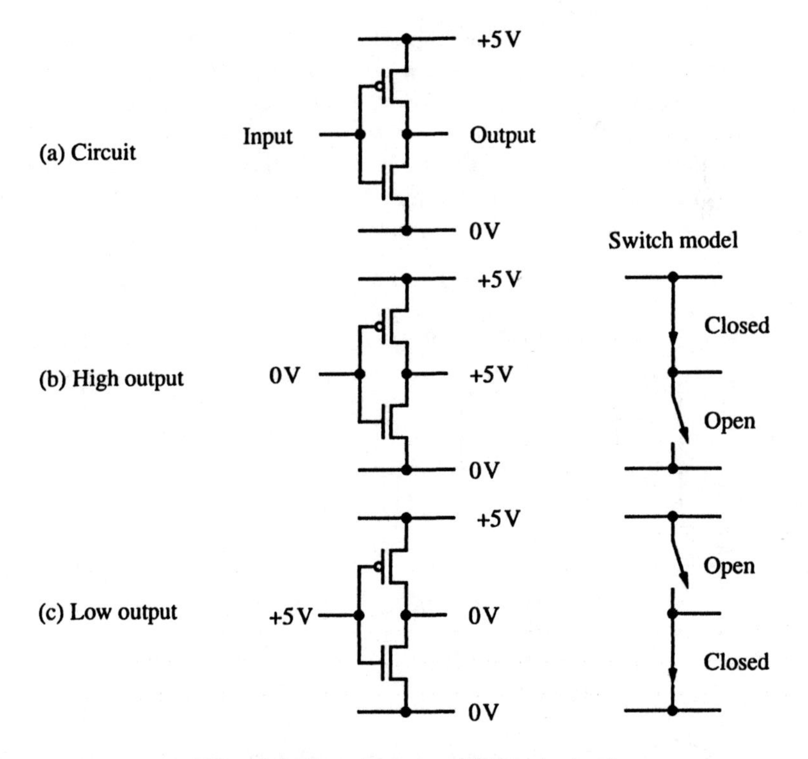

Figure 2.24 Complementary MOS logic circuit

1 and logic 0 values. Hence we have a circuit which can produce either a logic 0 (0 V) or a logic 1 (+5 V) dependent upon the input value. In this particular circuit, a logic 0 on the input creates a logic 1 on the output and a logic 1 on the input creates a logic 0 on the output.

Basic CMOS gates

The circuit shown in Figure 2.24 is actually a circuit of a NOT gate, and also forms the basis of NAND and NOR gates. A CMOS NAND gate is shown in Figure 2.25 with two inputs, *A* and *B*. The *n*-type transistors are connected in "series" to 0 V, and the *p*-type transistors are connected in "parallel" to +5 V. When *A* and *B* are both a 1, both *n*-type transistors are turned on causing the output to go to 0. Both *p*-type transistors will be off. When either *A* or *B* is a 0, the corresponding *p*-type transistor will be turned on causing the output to go to 1. Both *n*-type transistors will be off. A CMOS NOR gate is shown in Figure 2.26. In this case, the series and parallel configurations are in reverse positions. A CMOS AND gate can be obtained by using the NOR gate circuit with inverters on the inputs (as $AB = \overline{\overline{A} + \overline{B}}$). Similarly, a CMOS OR gate can be obtained by using the NAND gate circuit with inverters on the inputs (as $A + B = \overline{\overline{A}\,\overline{B}}$).

Combinational functions

CMOS circuits can be devised for arbitrary AND/OR functions using series and parallel transistor configurations. The *n*-type transistors form a network which defines the Boolean condition for a 0 since they pull down the output to a 0. The *p*-type transistors form a network which defines the Boolean conditions for a 1 since they pull up the output to a 1. The two Boolean conditions must be complementary. The pull-up network implements the true function, say F_{up}, and the pull-down network implements the complementary inverse function, say F_{down}.

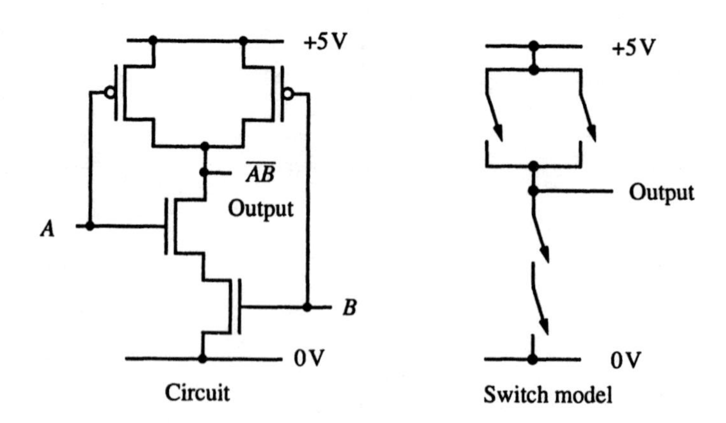

Figure 2.25 CMOS NAND gate

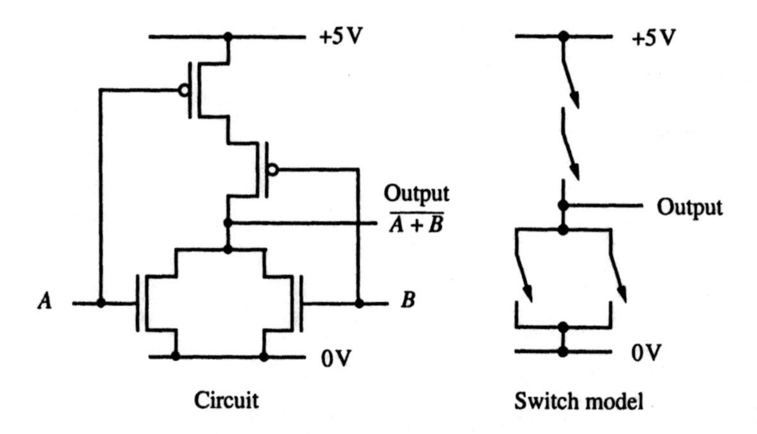

Figure 2.26 CMOS NOR gate

There are two basic configurations that can be used for each function, series or parallel. Series configurations form a connection when all transistors are switched "on" (fundamentally an AND operation). Parallel configurations form a connection when any transistor is switched "on" (fundamentally an OR operation). The F_{up} function is implemented with *p*-type transistors, and hence the complementary variables have to be applied (as noted by the bubbles on the transistors).

First let us look at the circuits for the NAND and NOR gates which we have already seen and see how F_{up} and F_{down} relate to these circuits. The conditions for the NAND gate circuit are:

Pull down to a 0 when $A = 1$ and $B = 1$
Pull up to a 1 when $A = 0$ or $B = 0$

or as two Boolean complementary functions, F_{down} and F_{up}:

$$F_{down} = AB$$
$$F_{up} = \bar{A} + \bar{B}$$

where:

$$F_{down} = \overline{F_{up}}$$

F_{up} is achieved using a parallel "OR" configuration and F_{down} a series "AND" configuration. F_{up} requires the signals A and B rather than $\bar{A}$ and $\bar{B}$, because of the use of *p*-type transistors. Hence we only need A and B to be applied to both the upper and lower transistors, see Figure 2.25.

For the NOR gate, the conditions are:

Pull down to a 0 when $A = 1$ or $B = 1$
Pull up to a 1 when $A = 0$ and $B = 0$

or as two Boolean functions, F_{down} and F_{up}:

$$F_{\text{down}} = A + B$$
$$F_{\text{up}} = \overline{A}\overline{B}$$

where again:

$$F_{\text{down}} = \overline{F}_{\text{up}}$$

Because of using *p*-type transistors, F_{up} requires A and B rather than $\overline{A}$ and $\overline{B}$, see Figure 2.26.

Arbitrary functions

Now let us look at an arbitrary function. Suppose we want a gate providing the Boolean function:

$$F = A\overline{B} + \overline{A}C$$

We establish the pull-up function as:

$$F_{\text{up}} = A\overline{B} + \overline{A}C$$

and the inverse (pull-down) function as:

$$F_{\text{down}} = \overline{A\overline{B} + \overline{A}C} = (\overline{A\overline{B}})(\overline{\overline{A}C})$$

$$= (\overline{A} + B)(A + \overline{C})$$

This leads to the circuit shown in Figure 2.27. Notice that now we need combinations of parallel and series configurations in both the upper structure and the lower structure. Also circuits are needed to invert the input variables (which are not shown). The implementation of the function, $f = A\overline{B} + \overline{A}C$, is *two-level* at the gate level, requiring two AND gates at the first level and a single OR gate at the second level. However, the function can be implemented in CMOS with a single "level" of transistors in Figure 2.27 (ignoring inverters).

SELF-ASSESSMENT QUESTION 2.13

What would happen if $F_{\text{down}} \neq F_{\text{up}}$?

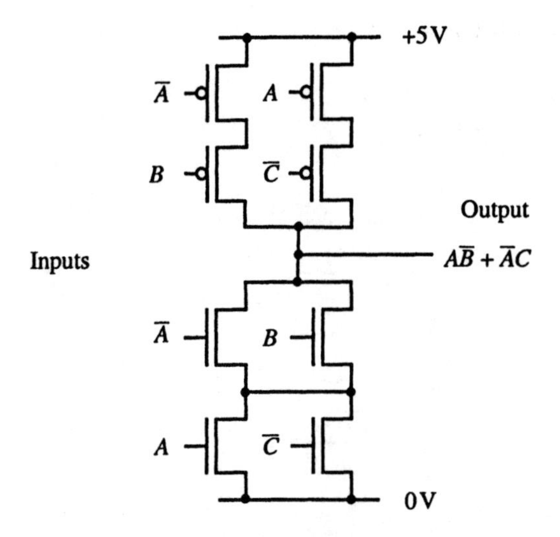

Figure 2.27 CMOS circuit for the function $A\bar{B} + \bar{A}C$

2.7 Summary

After this chapter, the reader should appreciate the following:

- The basic logic gates, AND, OR and NOT.
- The motives for the NAND and NOR gates.
- The functions of exclusive-OR/exclusive-NOR gates.
- Boolean identities and laws including the Principle of Duality and DeMorgan's theorem.
- The internal implementation of gates and functions in CMOS technology.

2.8 Tutorial questions

2.1 Prove the following using truth tables:

(a) $(A + B)(\bar{A} + C\bar{D}) = \bar{A}B + AC\bar{D}$
(b) $AB + \bar{A}C + BC = AB + \bar{A}C$

2.2 Under what conditions is the following Boolean equation valid?

$$A + BC = (A + B)C$$

2.3 Determine whether any of the following identities are valid:

(a) $XY + \overline{X}Z + YZ = XY + \overline{X}Z$

(b) $X \oplus Y = \overline{(\overline{\overline{X}Y}Z)(\overline{\overline{XY}Z})}$

2.4 Determine the value of X in each of the logic circuits shown in Figure 2.28.

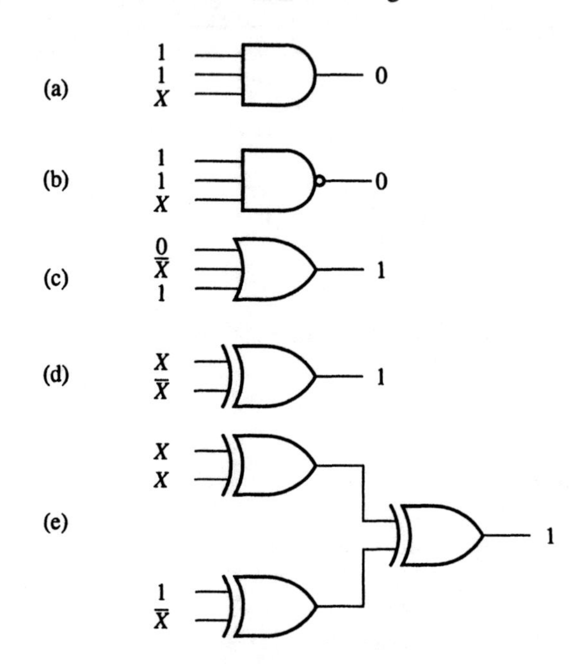

Figure 2.28 Circuits for Question 2.4

2.5 Determine the dual function of each of the following:

(a) $f(A, B, C) = A\overline{B} + \overline{A}B\overline{C}$
(b) $f(A, B, C) = \overline{ABC}(A + \overline{B}\,\overline{C})$

2.6 Write the dual equalities of the following:

(a) $A + \overline{A}B = A + B$
(b) $A + AB = A$

2.7 Reduce the following expressions using DeMorgan's theorem:

(a) $\overline{\overline{AB\overline{C}} + A\overline{B}C}$

(b) $(\bar{A} + B)(\overline{\overline{A + B} + C})$

2.8 If a circuit generates the function $f = (A + \bar{B})\bar{C}\,\bar{D}$ using positive logic representation, what does it generate in negative logic representation?

2.9 Prove the following basic identities for the exclusive-OR operation.

$$A \oplus B = A$$
$$A \oplus 1 = \bar{A}$$
$$A \oplus A = 0$$
$$A \oplus \bar{A} = 1$$
$$A \oplus \bar{B} = \bar{A} \oplus B = (\bar{A} \oplus \bar{B})$$

2.10 Design a CMOS circuit for each of the following functions:

(a) $f = ABC + \bar{A}\,\bar{C}$
(b) $f = (A + B)(\bar{A} + \bar{C})$

2.11 What Boolean function does the CMOS circuit shown in Figure 2.29 implement?

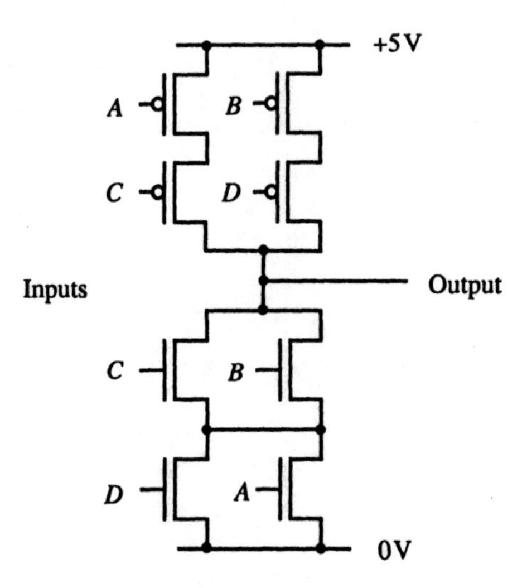

Figure 2.29 Circuit for Question 2.11

2.9 Suggested further reading

Wakerly devotes significant space to describing logic circuits including the various logic families (TTL, ECL, MOS, etc.):

Wakerly, J. F., *Digital Design Principles and Practices, 2nd edition*, Prentice Hall: Englewood Cliffs, New Jersey, 1994.

CHAPTER 3

Designing combinational circuits

Aims and objectives

In this chapter, we shall use the basic gates introduced in Chapter 2 to design simple logic circuits whose outputs are dependent upon the values placed on the inputs at that time. Such circuits are called combinational circuits because the output values depend upon particular combinations of input values. The gates in the previous chapter are basic combinational circuits themselves, and we shall use these gates to construct more complex combinational functions. Since it is often desired to implement a logic function with the minimum number of gates or minimum cost, we consider ways of rearranging and simplifying combinational logic circuits.

3.1 Combinational circuits

A combinational function can be defined by a Boolean expression. For example:

$$f_1 = AB + C$$

describes a function, f_1, which is a 1 when A and B are a 1 together or when C is a 1. The simplest way to implement this function is to use one AND gate to produce AB and one OR gate to produce $(AB) + C$, as shown in Figure 3.1. This combinational circuit has three inputs, A, B and C, and one output, f_1.

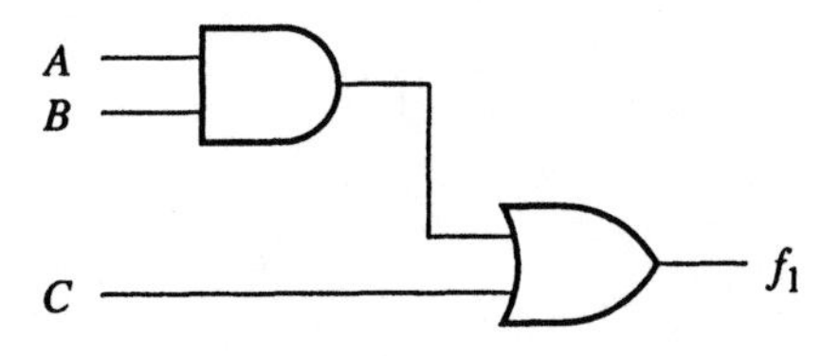

Figure 3.1 Gate implementation of function $f_1 = AB + C$

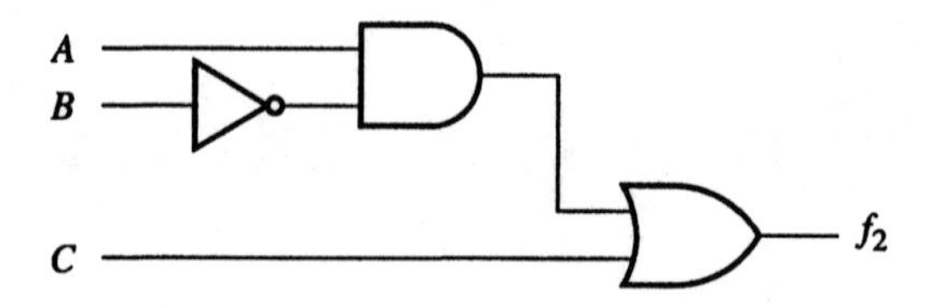

Figure 3.2 Gate implementation of function $f_2 = A\bar{B} + C$

A function could have variables which are complemented, for example:

$$f_2 = A\bar{B} + C$$

which describes a function which is a 1 when $A = 1$ and $B = 0$ together or when $C = 1$. The circuit implementation, assuming that the complement of B is not already available, requires a NOT gate as shown in Figure 3.2.

Figure 3.1 is a *two-level circuit* as signals pass through a maximum of two gates from input to output; Figure 3.2 is a three-level circuit including the NOT gate, though sometimes in counting levels we ignore the NOT gate so that this circuit could also be viewed as two-level. *Two-level circuits are significant because they implement AND–OR (and OR–AND) functions with the minimum signal delay from input to output.* We will look at alternative designs using more than two levels later.

Just as basic AND, OR and NOT functions can be described by a truth table, more complex functions can be described by a truth table. A truth table of f_1 is shown in Table 3.1 We can see from this table that f_1 is 1 in five instances, when $A = 0$, $B = 0$ and $C = 1$, when $A = 0$, $B = 1$ and $C = 1$, when $A = 1$, $B = 0$ and $C = 1$, when $A = 1$, $B = 1$ and $C = 0$, when $A = 1$, $B = 1$ and $C = 1$. Hence f_1 could have been written as:

$$f_1 = \bar{A}\bar{B}C + \bar{A}BC + A\bar{B}C + AB\bar{C} + ABC$$

Table 3.1 *Truth table of function* $f_1 = AB + C$

A	B	C	$AB + C$
0	0	0	0
0	0	1	1
0	1	0	0
0	1	1	1
1	0	0	0
1	0	1	1
1	1	0	1
1	1	1	1

each term corresponding to one of the entries in the truth table. If one of the terms computed to a 1, for example ABC when $A = 1, B = 1$ and $C = 1$, the whole expression is a 1.

SELF-ASSESSMENT QUESTION 3.1

Draw the truth table of the function f_2.

Each function is obtained from the truth table by taking each entry as one term of the function. This approach may create a function which has more terms than necessary and hence would require more gates or more inputs on each gate (or both). Clearly we want the expression which leads to the least expensive implementation. Simplifying expressions for this goal will be described later, but first let us look at some algebraic properties of expressions.

3.2 Implementing Boolean expressions

3.2.1 Sum-of-product expressions

Expression can of course be written in many ways. However there is a "standard" form which appears very often, an expression consisting of Boolean variables connected with AND operators to form terms that are connected with OR operators. The expressions in the previous section were of this form. AND operations can be viewed as Boolean multiplication (product). A term consisting of variables connected with AND operators is a *product term.* OR operations can be viewed as Boolean summation. Hence an expression with product terms connected by OR operators is a *sum-of-product (SOP) expression.* The following is a sum-of product expression:

$$f(A, B, C) = A\overline{B} + \overline{A}B\overline{C} + AC$$

Here the variables in the expression A, B and C are indicated on the left-hand side of the function. Sum-of-product expressions can be implemented using AND gates and an OR gate, as shown in Figure 3.3.

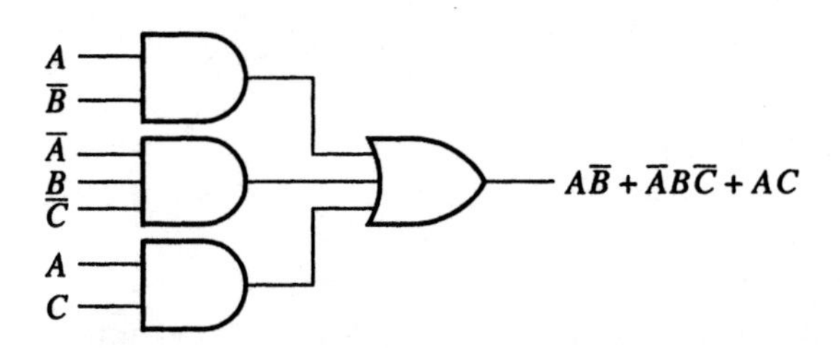

Figure 3.3 Two-level implementation of the function $f = A\overline{B} + \overline{A}B\overline{C} + AC$

Canonical expressions

In the previous example, not all the variables appear in each product term. If a sum-of-product expression contains all the variables (or their complements) in each product term, such as in:

$$f(A, B, C) = \bar{A}\bar{B}\bar{C} + \bar{A}B\bar{C} + \bar{A}BC + A\bar{B}C$$

the expression is called a *canonical sum-of-product expression*. The product terms in a canonical sum-of-product expressions are called *minterms*.

SELF-ASSESSMENT QUESTION 3.2

Draw the logic circuit for the function $f(A, B, C) = \bar{A}\bar{B}\bar{C} + \bar{A}B\bar{C} + \bar{A}BC + A\bar{B}C$

Each minterm in a canonical expression can be described by a binary number such that each digit in the number is a 1 if the variable in the minterm is true and a 0 if the variable is complemented. Hence there is a unique binary number for each minterm, i.e. $\bar{A}\bar{B}\bar{C} = 000 = P_0$, $\bar{A}\bar{B}C = 001 = P_1$, etc., as shown in Table 3.2.

Table 3.2 *Minterm numbering*

Minterm	*Number*	*Notation*
$\bar{A}\bar{B}\bar{C}$	000	P_0
$\bar{A}\bar{B}C$	001	P_1
$\bar{A}B\bar{C}$	010	P_2
$\bar{A}BC$	011	P_3
$A\bar{B}\bar{C}$	100	P_4
$A\bar{B}C$	101	P_5
$AB\bar{C}$	110	P_6
ABC	111	P_7

A canonical sum-of-product expression can be described using the standard algebraic symbol for summation, Σ, i.e.:

$$f(A, B, C) = \bar{A}\bar{B}\bar{C} + \bar{A}B\bar{C} + \bar{A}BC + A\bar{B}C = P_0 + P_2 + P_3 + P_5 = \Sigma(0, 2, 3, 5)$$

SELF-ASSESSMENT QUESTION 3.3

Write the logic expression for the function $f(A, B, C) = \Sigma(0, 3, 7)$.

Truth table description

Design problems might begin with a truth table description of the circuit. The truth table description naturally provides a canonical sum-of-product expression since each entry in the truth table corresponds to one minterm. Table 3.3 shows the truth

Table 3.3 *Truth table of the function*
$f(A, B, C) = \bar{A}\bar{B}\bar{C} + \bar{A}B\bar{C} + \bar{A}BC + A\bar{B}C$

A	B	C	f	
0	0	0	1	← $\bar{A}\bar{B}\bar{C}$
0	0	1	0	
0	1	0	1	← $\bar{A}B\bar{C}$
0	1	1	1	← $\bar{A}BC$
1	0	0	0	
1	0	1	1	← $A\bar{B}C$
1	1	0	0	
1	1	1	0	

table of the function $f(A, B, C) = \bar{A}\bar{B}\bar{C} + \bar{A}B\bar{C} + \bar{A}BC + A\bar{B}C$. Each 1 in the output column corresponds to one minterm.

Conversion into canonical form

A canonical sum-of-product expression can be obtained from a non-canonical sum-of-product expression by multiplying each term by $(X + \bar{X})(Y + \bar{Y})(\ ...\ )$ where X, Y, etc. are variables not in the product term. The expression is then expanded.

EXAMPLE

The non-canonical sum-of-product expression:

$$f(A, B, C) = A\bar{B} + \bar{A}C + A\bar{B}C$$

expands to the canonical sum-of-product expression:

$$\begin{aligned} f(A, B, C) &= A\bar{B}(C + \bar{C}) + \bar{A}C(B + \bar{B}) + A\bar{B}C \\ &= A\bar{B}C + A\bar{B}\bar{C} + \bar{A}BC + \bar{A}\bar{B}C + A\bar{B}C \\ &= A\bar{B}C + A\bar{B}\bar{C} + \bar{A}BC + \bar{A}\bar{B}C \end{aligned}$$

Duplicated terms such as the second $A\bar{B}C$ above are removed ($A\bar{B}C + A\bar{B}C = A\bar{B}C$ from the basic identity $A + A = A$).

SELF-ASSESSMENT QUESTION 3.4

How could the function $f(A, B, C) = A$ be converted into canonical form?

Canonical expressions can be converted into simpler non-canonical expressions by repeated factorization, removing $(X + \bar{X})$ where X appears in one term and $\bar{X}$ appears in the other term. This process, as we shall see, is the basis of techniques to simplify Boolean expressions.

3.2.2 Product-of-sum expressions

Though not as common, another standard form of expression is with variables (or their complement) connected with OR operators to form terms that can be connected with AND operators. A term consisting of variables connected with OR operators is a *sum term*. The expression with sum terms connected with OR operators is a *product-of-sum expression* which can be canonical if each variable (or its complement) appears in each term. The following is a (non-canonical) product-of-sum expression:

$$f(A, B, C) = (\bar{A} + B + \bar{C})(A + \bar{B})(A + C)$$

whereas:

$$f(A, B, C) = (\bar{A} + \bar{B} + \bar{C})(\bar{A} + \bar{B} + C)(A + B + C)$$

is canonical. The sum terms in a canonical product-of-sum expression are called *maxterms*.

Product-of-sum expressions can be implemented using OR gates and an AND gate, as shown in Figure 3.4 (for the first expression above).

SELF-ASSESSMENT QUESTION 3.5

Draw the logic circuit for $f(A,B,C) = (\bar{A}+\bar{B}+\bar{C})(\bar{A}+\bar{B}+C)(\bar{A}+B+C)(A+B+C)$.

Just as minterms can be described by binary numbers, each maxterm in a canonical expression can be described by a binary number such that each digit in the number is a 1 if the variable in the maxterm is true and a 0 if the variable is complemented, for example the maxterm, $A + \bar{B} + C = 101 = S_5$. Maxterm numbering is shown in Table 3.4. A canonical product-of-sum expression can be described using the standard algebraic symbol for products, Π, i.e.:

$$f(A, B, C) = (\bar{A} + B + \bar{C})(A + \bar{B} + C)(A + B + \bar{C}) = S_2 S_5 S_6 = \Pi(2, 5, 6)$$

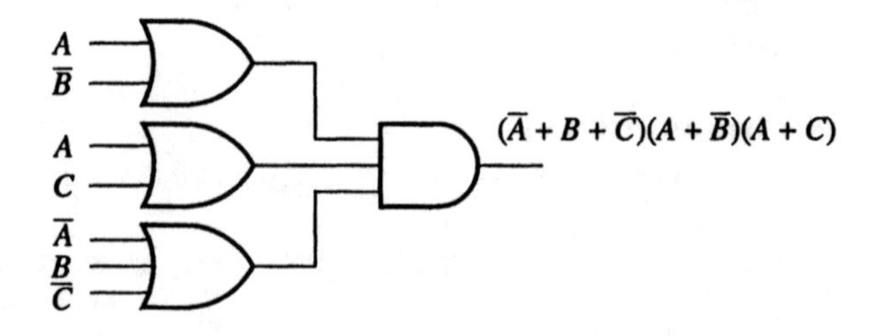

Figure 3.4 Two-level implementation of the function $f = (\bar{A} + B + \bar{C})(A + \bar{B})(A + C)$

Table 3.4 *Maxterm numbering*

Minterm	*Number*	*Notation*
$\bar{A}+\bar{B}+\bar{C}$	000	S_0
$\bar{A}+\bar{B}+C$	001	S_1
$\bar{A}+B+\bar{C}$	010	S_2
$\bar{A}+B+C$	011	S_3
$A+\bar{B}+\bar{C}$	100	S_4
$A+\bar{B}+C$	101	S_5
$A+B+\bar{C}$	110	S_6
$A+B+C$	111	S_7

SELF-ASSESSMENT QUESTION 3.6

Write the logic expression for the function $f(A, B, C) = \Pi(0, 3, 7)$.

Conversion into canonical form

The dual method of converting sum-of-product expressions, i.e. using $X\bar{X} = 0$, cannot be used to convert a non-canonical product-of-sum expression into a canonical product-of-sum expression. A method to convert a non-canonical product-of-sum expression into a canonical product-of-sum expression is to form the inverse function, $\bar{f}$, using DeMorgan's theorem. The resulting expression will be in sum-of-product form. Then apply the procedure to convert a sum-of-product expression to canonical form. Then apply DeMorgan's theorem to convert the expression into product-of-sum form, which will be canonical.

EXAMPLE

$$
\begin{aligned}
f(A, B, C) &= (A+\bar{B}+C)(\bar{A}+B) \\
\bar{f}(A, B, C) &= \overline{(A+\bar{B}+C)(\bar{A}+B)} \\
&= \overline{(A+\bar{B}+C)} + \overline{(\bar{A}+B)} \\
&= \bar{A}B\bar{C} + A\bar{B} \\
&= \bar{A}B\bar{C} + A\bar{B}(C+\bar{C}) \\
&= \bar{A}B\bar{C} + A\bar{B}C + A\bar{B}\bar{C} \\
f(A, B, C) &= \overline{\overline{\bar{A}B\bar{C} + A\bar{B}C + A\bar{B}\bar{C}}} \\
&= (\overline{\bar{A}B\bar{C}})(\overline{A\bar{B}C})(\overline{A\bar{B}\bar{C}}) \\
&= (A+\bar{B}+C)(\bar{A}+B+\bar{C})(\bar{A}+B+C)
\end{aligned}
$$

The minterm and maxterm notations can be used to develop a more formal conversion method (see Wilkinson, 1996, for details).

3.3 Alternative implementations

Boolean functions can be manipulated in much the same way as conventional algebraic expressions; expressions can be factorized, parentheses inserted to indicate precedence, etc. Different variations may suggest particular implementations.

Sum-of-product expressions

As we have seen, sum-of-product expressions suggest the use of AND gates to generate the product terms, and an OR gate to generate the sum. The sum-of-product expression:

$$f(A, B, C) = \bar{A}\bar{B}\bar{C} + \bar{A}BC + A\bar{B}C$$

by observation, can be implemented using the AND–OR gate circuit shown in Figure 3.5. If we apply DeMorgan's theorem to f, we can obtain:

$$f(A, B, C) = \overline{(\overline{\bar{A}\bar{B}\bar{C}})(\overline{\bar{A}BC})(\overline{A\bar{B}C})}$$

for the NAND–NAND gate implementation shown in Figure 3.6. Figure 3.6 shows an interesting and very convenient characteristic of NAND gate implementation; the NAND gates simply replace the AND and OR gates in the AND–OR gate implementation of Figure 3.5. The first row of NAND gates actually perform the AND operation logically since this is the operation specified in the original equation. The final NAND gate collecting the outputs from the first row is performing the OR operation logically. The actual output signal from each of the first row of gates is a logic 0 (low) when the all the inputs to the gate are a logic 1 (rather than a logic 1 level). Any of these 0's will cause the output f to become a 1.

To identify the intended function more clearly, an alternative symbol exits for NAND gates as shown in Figure 3.7. Notice the use of inversion circles on the inputs of a conventional OR gate symbol. Hence the function is $\bar{X} + \bar{Y} + \bar{Z}$ (the same as a NAND function, $\overline{XYZ}$) where X, Y, and Z are the inputs to the gate. We could view

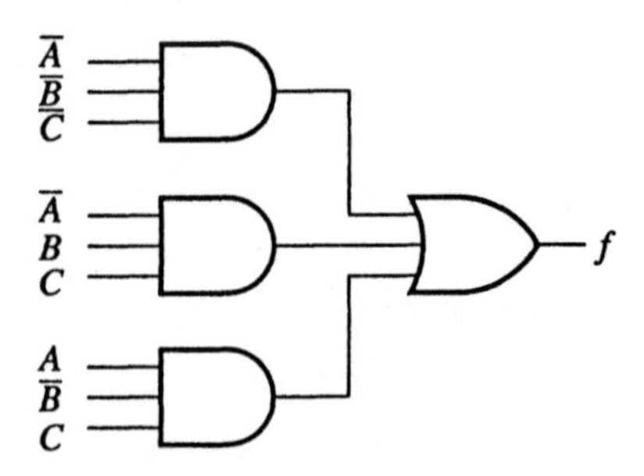

Figure 3.5 AND–OR implementation of the function $f = \bar{A}\bar{B}\bar{C} + \bar{A}BC + A\bar{B}C$

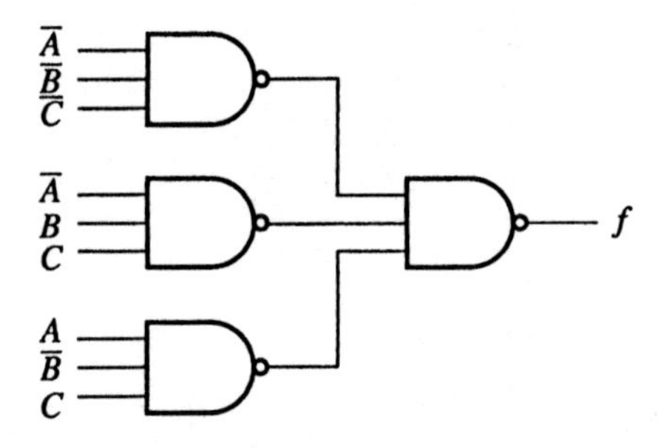

Figure 3.6 NAND–NAND implementation of the function $f = \bar{A}\bar{B}\bar{C} + \bar{A}BC + A\bar{B}C$

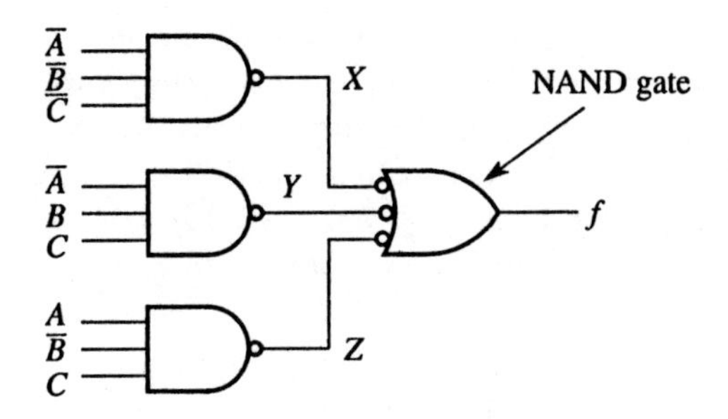

Figure 3.7 NAND–NAND implementation of the function $f = \bar{A}\bar{B}\bar{C} + \bar{A}BC + A\bar{B}C$ *using an alternative symbol*

the inversion circles on the output the NAND gates and on the input of the alternative symbol as cancelling in Figure 3.7. The alternative symbol would be used in situations where we wish to indicate the OR operation, as in this case. If an AND gate were used for the output gate in Figure 3.7, an additional output inversion circle would be present in the symbol.

Product-of-sum expressions

A product-of-sum expression can be implemented using OR gates to generate the sum terms and an AND gate to generate the product. For example, the expression:

$$f(A, B, C) = (\bar{A} + \bar{B} + \bar{C})(\bar{A} + B + C)(A + \bar{B} + C)$$

by observation, can be implemented using the gate circuit shown in Figure 3.8. If we apply DeMorgan's theorem to f, we can obtain:

$$f(A, B, C) = \overline{\overline{(\bar{A} + \bar{B} + \bar{C})} + \overline{(\bar{A} + B + C)} + \overline{(A + \bar{B} + C)}}$$

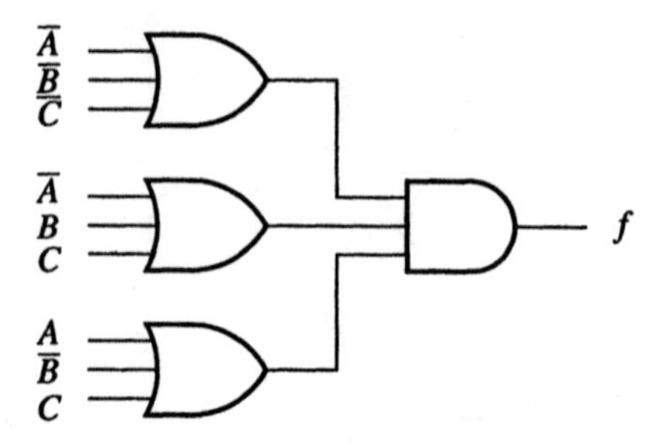

Figure 3.8 OR–AND implementation of the function $f = (\overline{A} + \overline{B} + \overline{C})(\overline{A} + B + C)(A + \overline{B} + C)$

for a NOR–NOR gate implementation shown in Figure 3.9. Again there is a very convenient circuit characteristic; the NOR gates simply replace the AND and OR gates in the OR–AND gate implementation directly. An alternative symbol could be used to highlight the Boolean operations intended as shown in Figure 3.10. In this case, we wish to indicate a final AND operation. In this symbol, the inversion circles are placed on the inputs of an AND symbol.

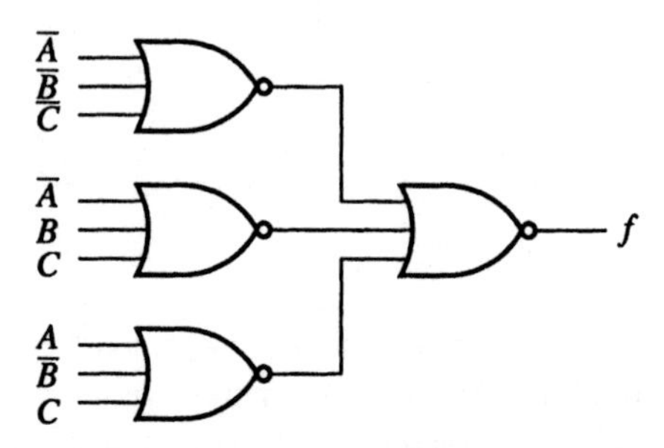

Figure 3.9 NOR–NOR implementation of the function $f = (\overline{A} + \overline{B} + \overline{C})(\overline{A} + B + C)(A + \overline{B} + C)$

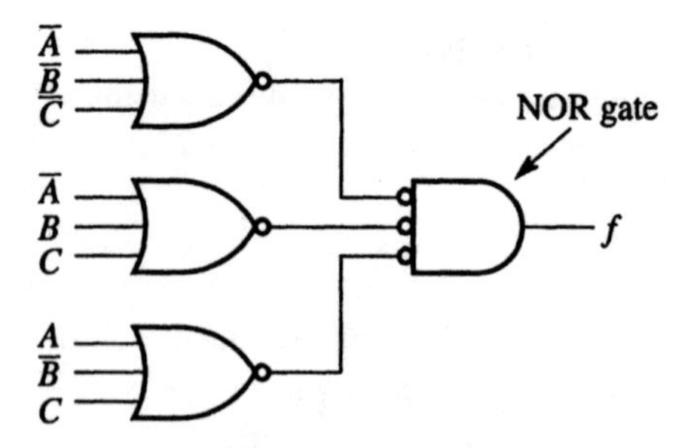

Figure 3.10 NOR–NOR implementation of the function $f = (\overline{A} + \overline{B} + \overline{C})(\overline{A} + B + C)(A + \overline{B} + C)$*, using an alternative symbol*

Multilevel implementations

In all our implementations so far, the inverters necessary to generate the complements are not shown. Disregarding these inverters, the circuits are two-level, that is, the signals pass through two gates from input to output. Multilevel implementations can be achieved through factorizing the expressions, which may reduce the number of inputs to individual gates but at the expense of a greater delay.

EXAMPLE

To implement the function:

$$f = ABC + ABD + ABE + AF$$

we could factorize the expression into:

$$f = AB(C + D + E) + AF$$

leading to the circuit shown in Figure 3.11.

SELF-ASSESSMENT QUESTION 3.7

Factorize the function $f = ABC + ABD + ABE + AF$ in a different way to the example.

3.4 Simplifying logic circuits

Clearly we want the most cost-effective circuit which meets a design specification (system function, gate delay, speed, cost, etc.). If designing using packaged gates, usually we want to minimize the number of packages, taking into account the overall gate delay, and available gates within the packages. If designing in VLSI (very large-scale integration), we may want to minimize the number of transistors.

Suppose we intend to design using packaged gates. There are several ways to

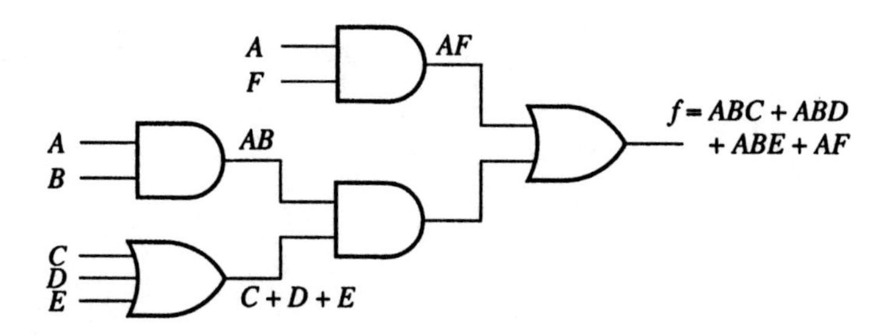

Figure 3.11 Implementing the function f = ABC + ABD + ABE + AF through factorization

simplify a circuit. These methods start with the Boolean expression describing the function. Simplifying the expression usually means reducing the number of variables in each term and reducing the number of terms. (Of course, these may be conflicting requirements.) The term *minimization* is used to describe the simplification process. In a practical logic design, the purpose of minimization is to obtain an expression which will cost less or operate faster than by implementing the original expression. In some instances, less levels is desirable even though more gates may be need as the signal delay through the use of a logic circuit is directly linked to the number of levels of gates.

3.4.1 Simplifying Boolean expressions algebraically

Algebraic simplification applies Boolean identities and laws. The following algebraic strategies can be employed, dependent upon the actual Boolean expression. In the most part, we have to identify the best method by intuition. More formal methods are described later.

1. Applying identities after grouping

In this strategy, the associative law is used to group terms, and identities such as $A + 1 = 1$ are then applied to grouped terms to reduce the expression.

EXAMPLE

$$\begin{aligned} f(A, B, C) &= A + A\overline{C} + ABC \\ &= A(1 + \overline{C} + BC) \\ &= A \end{aligned}$$

Terms can be duplicated if necessary to create further groups in an expression (because $A + A = A$).

An identity not usually regarded as a basic identity, but of use in simplifying equations is:

$$A + \overline{A}B = A + B$$

which can be proved easily by a truth table. This identity can be applied in various situations.

EXAMPLE

In the following, the identity $A + \overline{A}B = A + B$ is used twice:

$$\begin{aligned} f(A, B, C) &= A + \overline{B} + \overline{A}BCD \\ &= A + \overline{A}BCD + \overline{B} \\ &= A + BCD + \overline{B} \\ &= A + \overline{B} + CD \end{aligned}$$

Consensus theorem

Another identity in the same general category as $A + \bar{A}B = A + B$ is:

$$AB + \bar{A}C + BC = AB + \bar{A}C$$

Here BC is redundant and can be removed. The identity can also be proved by a truth table. Actually this identity with its dual:

$$(A + B)(\bar{A} + C)(B + C) = (A + B)(\bar{A} + C)$$

forms what is called the *Consensus theorem.* The redundant term is the consensus of the other two terms.

SELF-ASSESSMENT QUESTION 3.8

Simplify:

$$f = ABC + \bar{A}DE + BCDE$$

using the Consensus theorem. (Clue: group together BD and DE.)

2. Expansion into canonical form followed by reduction

We have seen the use of multiplying terms of a sum-of-product expression by $X + \bar{X}$ to get a canonical expression. The canonical expression provides terms which can be recombined to obtain a simpler expression. First the expression is expanded into canonical form. Terms are then grouped together to reduce the expression. Terms can be duplicated if necessary to form groups.

EXAMPLE

$$\begin{aligned} f(A, B, C) &= A\bar{B} + AB\bar{C} + \bar{A}B\bar{C} + \bar{A}BC \\ &= A\bar{B}(C + \bar{C}) + AB\bar{C} + \bar{A}B\bar{C} + \bar{A}BC \\ &= A\bar{B}C + A\bar{B}\bar{C} + AB\bar{C} + \bar{A}B\bar{C} + \bar{A}BC \\ &= A\bar{B}(C + \bar{C}) + A\bar{C}(B + \bar{B}) + \bar{A}B(C + \bar{C}) \\ &= A\bar{B} + A\bar{C} + \bar{A}B \end{aligned}$$

This method is guaranteed to minimize a sum-of-product expression. A product-of-sum expression can be minimized by first obtaining the inverse function as a sum-of-product function. This function is then minimized and changed back to the true function using DeMorgan's theorem.

SELF-ASSESSMENT QUESTION 3.9

What other way could the function $f(A, B, C) = A\bar{B} + AB\bar{C} + \bar{A}B\bar{C} + \bar{A}BC$ be reduced?

3. By using DeMorgan's theorem

DeMorgan's theorem can be used to reduce expressions. Generally DeMorgan's theorem is most useful with complex expressions involving multiple levels of inversion. Repeated use of DeMorgan's theorem can reduce the number of levels.

An aid to remembering how to apply DeMorgan's theorem is as follows: If there is a continuous bar over an expression consisting of terms linked with an operator (AND or OR), *break-up the bar* into parts and change the operator (from AND to OR, or from OR to AND). Similarly, if there are separate bars over terms linked by an operator, *combine the bars* and change the operator.

EXAMPLE

To simplify:

$$f(A, B, C) = \overline{(ABD + \overline{\overline{B\bar{C}D}})\bar{B}C + \bar{A}C}$$

we could first "break-up" the bar to get:

$$= \left(\overline{(ABD + \overline{\overline{B\bar{C}D}})\bar{B}C}\right)(\overline{\bar{A}C})$$

Continuing this strategy of "breaking-up bars", we get:

$$= \left(\overline{ABD + (\overline{\overline{B\bar{C}D}})} + (\overline{\bar{B}C})\right)(A + \bar{C})$$

$$= ((\overline{ABD})(B\bar{C}D) + (B + \bar{C}))(A + \bar{C})$$

$$= ((\bar{A} + \bar{B} + \bar{D})(B\bar{C}D) + B + \bar{C})(A + \bar{C})$$

Multiplying out (and noting that $B\bar{B} = 0$ and $D\bar{D} = 0$), we get:

$$= (\bar{A}B\bar{C}D + B + \bar{C})(A + \bar{C})$$

Continuing (noting also that $\bar{C}\bar{C} = \bar{C}$):

$$= AB + A\bar{C} + \bar{A}B\bar{C}D + B\bar{C} + \bar{C}$$

Regrouping, we get:

$$= AB + \bar{C}(A + \bar{A}BD + B + 1)$$

$$= AB + \bar{C}$$

DeMorgan's theorem can be applied across various terms and in different ways in most expressions. For example, there may be a choice between "combining bars" and "breaking-up bars", and one will need to decide how to proceed. The theorem in itself is not usually sufficient to minimize an equation; basic Boolean identities are usually also needed.

SELF-ASSESSMENT QUESTION 3.10

"Break-up the bars" in the expression $f = \overline{A + BC}$.

SELF-ASSESSMENT QUESTION 3.11

"Combine the bars" in the expression $\bar{A} + \bar{B}\bar{C}$.

3.4.2 Karnaugh map minimization method

Though algebraic minimization is capable of reducing any expression to its simplest form (or to any other desired form), it relies on the skill of the individual in applying the appropriate rules. Given a canonical sum-of-product expression, we have seen that one rule can be used repeatedly to reduce any sum-of-product expression, namely $PA + P\bar{A} = P(A + \bar{A}) = P$ where P is any product term and A is a variable that appears in the terms. This is described in the algebraic method 2 above. However, it also relies on being able to recognize the terms that can be combined. The *Karnaugh map method* provides a simple visual method of identifying the terms that can be combined.

A *Karnaugh map* consists of a two-dimensional array of squares. Each square corresponds to one minterm. The minterms of adjacent squares, either horizontally or vertically, have the same variables except that in one minterm one variable is true and in the other the same variable is complemented. Squares are marked with a 1 for those minterms in the function to be minimized. Adjacent 1's, either horizontally or vertically, can be combined into a group corresponding to a single product term, eliminating the variable that was different (using $(P(A + \bar{A}) = P)$. When two minterms are combined into a simpler product term, the group can be further combined with similarly sized adjacent groups as we shall see.

A three-variable Karnaugh map is shown in Figure 3.12. There are eight combinations of three variables and each combination is assigned a square on the Karnaugh

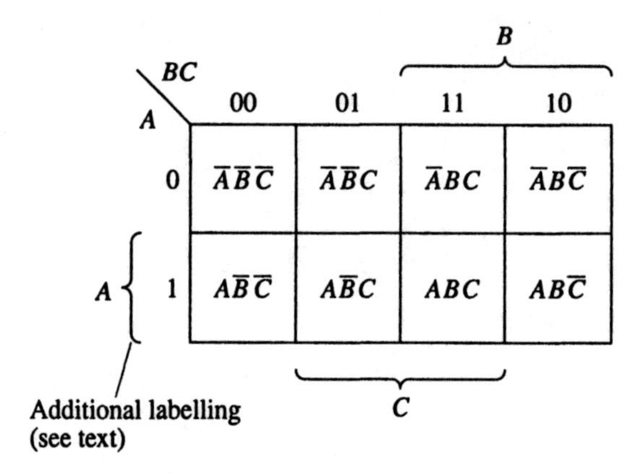

Figure 3.12 Three-variable Karnaugh map

C \ AB	00	01	11	10
0	$\bar{A}\bar{B}\bar{C}$	$\bar{A}B\bar{C}$	$AB\bar{C}$	$A\bar{B}\bar{C}$
1	$\bar{A}\bar{B}C$	$\bar{A}BC$	ABC	$A\bar{B}C$

Figure 3.13 Three-variable Karnaugh map with alternative ordering of variables

map. Squares are labelled by the three variables along two edges of the map. A 0 indicates that the variable is complemented and a 1 indicates that the variable is not complemented. Adjacent squares are assigned terms which differ by one variable. Hence we have 00, 01, 11, 10 across the top. As we move from one square to the next horizontally, one variable changes. This sequence is actually a code called *Gray code*. The vertical sequence is the same. In this particular case, there is only one variable vertically. A square is identified by the vertical and horizontal variables, the first square uppermost on the left being $\bar{A}\,\bar{B}\,\bar{C}$ (i.e. 000).

An alternative form of labelling is also shown in Figure 3.12; parentheses extending along the edges where the variable is true. This form of labelling is useful for identifying the squares where a variable is true.

The order of the variables can be different, for example C vertically and AB horizontally, as shown in Figure 3.13. We shall use the first order of variables because it is easier to find the square from the term by reading left to right in Figure 3.12, as given in the term, i.e. ABC is A down and BC across. However, the order shown in Figure 3.13 is also used, and it is personal preference how the variables are ordered.

A 1 mark on the Karnaugh map will also indicate when the function is a 1. The remaining squares can be marked with 0's to indicate that the function will be a 0 in these instances. Often the 0's are not given, though they are here. An example of mapping a function, $f = \bar{A}\,\bar{B}\,\bar{C} + \bar{A}BC + ABC$, is shown in Figure 3.14.

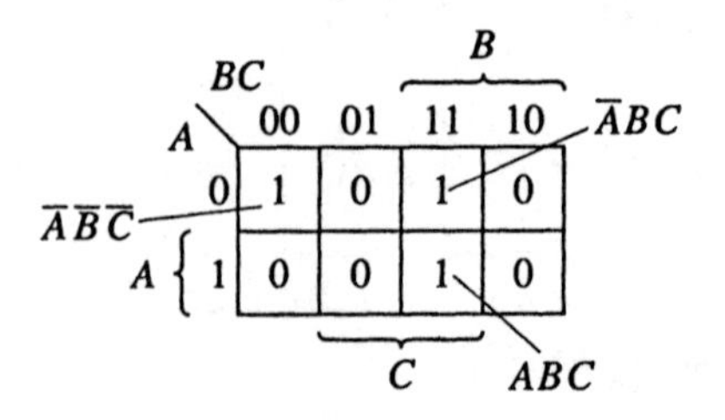

Figure 3.14 Mapping the function f(A, B, C) = $\bar{A}\bar{B}\bar{C} + \bar{A}BC + ABC$ on a three-variable Karnaugh map

SELF-ASSESSMENT QUESTION 3.12

Map the function $f = A\overline{B}C + AB\overline{C} + \overline{A}B\overline{C}$ on a Karnaugh map.

Minimization technique

The minimization technique usually begins with the sum-of-product expression to be minimized. In the first instance, let us assume that this is a canonical expression. There are three basic steps:

1. Mark those squares with a 1 that correspond to the terms in the expression.
2. Form the 1's into the largest valid groups of 1's.
3. Read the terms from the map which correspond to the groups ("cover" the 1's).

EXAMPLE

Suppose the expression to minimize is:

$$f(A, B, C) = \overline{A}\,\overline{B}\,\overline{C} + \overline{A}\,\overline{B}C + AB\overline{C}$$

The Karnaugh map of this function is shown in Figure 3.15. There are three terms and hence three 1's on the map. Two of these 1's are adjacent and indicate that the two terms can be combined, i.e. $\overline{A}\,\overline{B}\,\overline{C}$ and $\overline{A}\,\overline{B}C$ can be combined into $\overline{A}\,\overline{B}$. The final 1 is not adjacent to any other 1 and cannot be combined. The combined group is shown by circling the terms together and labelling the group. The variable being eliminated can be recognized by looking at the labels. In one square the variable will be a 0 and in the adjacent square the variable will be a 1. The alternative labelling using parentheses is useful when trying to find the common term. In this labelling, the bottom row all have A true, the right two columns all have B true and the central two columns have C true. The final minimized expression is $\overline{A}\,\overline{B} + AB\overline{C}$.

Groups between edges

Squares on one edge are also considered adjacent to squares on the opposite edge (top and bottom or left and right) as such squares correspond to terms with one variable different.

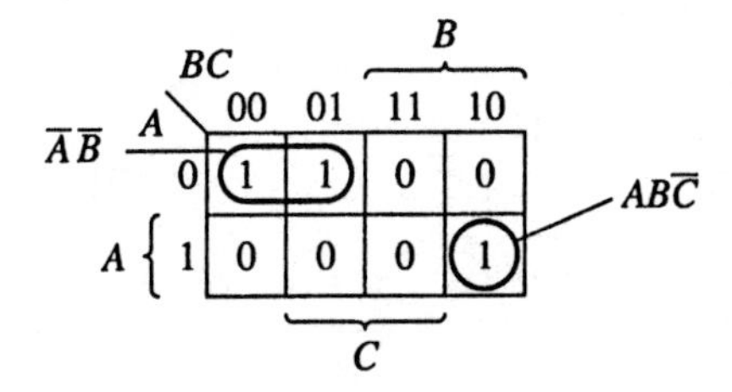

Figure 3.15 Three-variable Karnaugh map of function $f(A, B, C) = \overline{A}\,\overline{B}\,\overline{C} + \overline{A}\,\overline{B}C + AB\overline{C}$

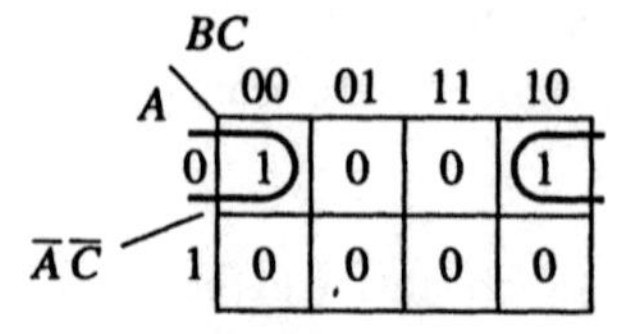

Figure 3.16 Wraparound characteristic of Karnaugh maps

EXAMPLE

Figure 3.16 shows an example of this sort of grouping. This particular minimization corresponds to:

$$f(A, B, C) = \bar{A}\,\bar{B}\,\bar{C} + \bar{A}B\bar{C} = \bar{A}\,\bar{C}(B + \bar{B}) = \bar{A}\,\bar{C}$$

Larger groups

So far, we have seen how to combine two terms into one group to eliminate one variable. If two such combined groups have only one variable different (and will be adjacent), these groups could be combined again to eliminate a second variable. The larger groups can be formed in the same way as the smaller groups by looking for groups that are adjacent, either horizontally or vertically. Examples of larger groups are shown in Figure 3.17. In each case, the final minimized term is given by those variables which are common to all the squares in the group. A group of two squares eliminate one variable, a group of four squares eliminates two variables and a group of eight squares eliminates three variables. Groups can use edges, for example f_3 in Figure 3.17(c). In Figure 3.17(d), f_4 would result in a constant 1 irrespective of the variable values. Notice in all cases, only groups of 2, 4, 8, ..., 2^n squares can be used

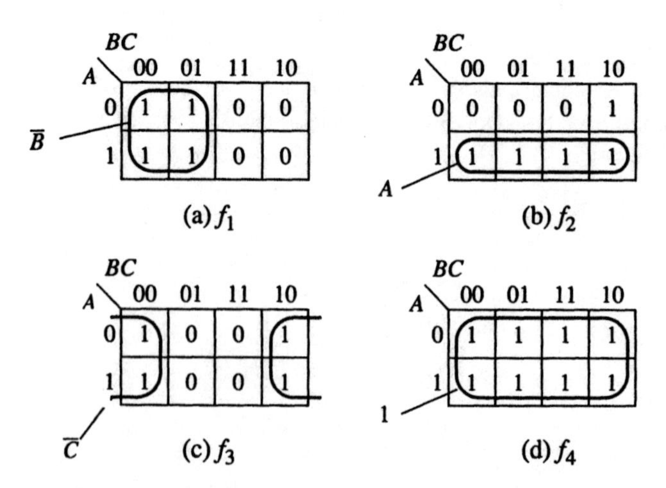

Figure 3.17 Examples of larger groups in Karnaugh maps

in a Karnaugh map for minimization and the groups must form a square or a rectangle. The number of squares on each side of the group must also be a power of 2.

Four-variable maps

Functions of four variables requires a Karnaugh map with sixteen squares, as shown in Figure 3.18. The minimization technique is the same as for smaller Karnaugh maps. Adjacent 1's are grouped together and the largest possible groups are formed. With a four-variable map, which has sixteen squares, we have the prospect of groups of two 1's, four 1's, eight 1's or even all sixteen 1's (when all the squares have 1's resulting in the function being a 1 permanently).

EXAMPLE

Suppose the function:

$$f(A,B,C,D) = \bar{A}\bar{B}\bar{C}D + \bar{A}\bar{B}CD + \bar{A}B\bar{C}D + \bar{A}BCD + AB\bar{C}D + A\bar{B}\bar{C}\bar{D} + A\bar{B}\bar{C}D + A\bar{B}C\bar{D}$$

is to be minimized. The function "mapped" onto a Karnaugh map is shown in Figure 3.19 resulting in the minimized function:

$$f(A,B,C,D) = \bar{A}D + \bar{C}D + A\bar{B}\bar{D}$$

The previous example shows that groups can be selected that overlap. This is done to create the largest possible groups and hence the smallest number of variables in the terms.

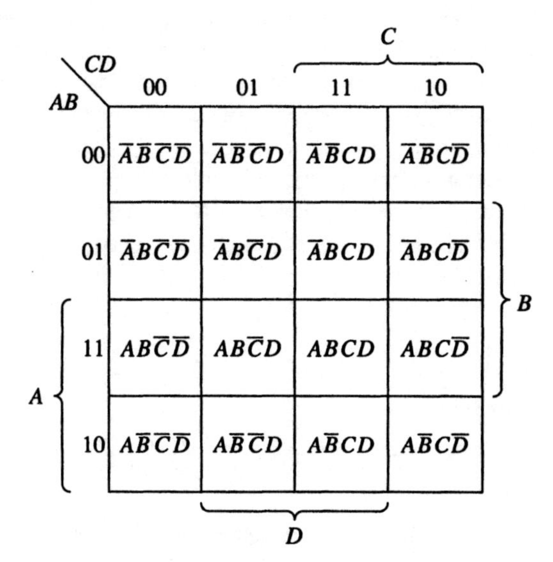

Figure 3.18 Four-variable Karnaugh map

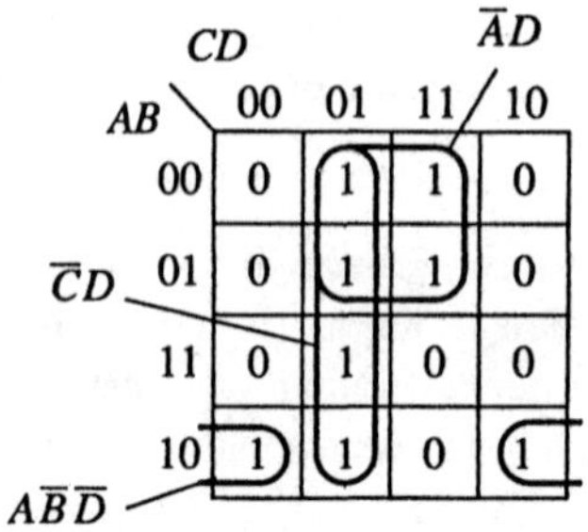

$\bar{A}D$

AB \ CD	00	01	11	10
00	0	1	1	0
01	0	1	1	0
11	0	1	0	0
10	1	1	0	1

$\bar{C}D$

$A\bar{B}\bar{D}$

Figure 3.19 $f(A, B, C, D) = \bar{A}\bar{B}\bar{C}D + \bar{A}\bar{B}CD + \bar{A}B\bar{C}D + \bar{A}BCD + AB\bar{C}D + A\bar{B}\bar{C}\bar{D} + A\bar{B}\bar{C}D + A\bar{B}C\bar{D}$ *mapped onto a four-variable Karnaugh map*

There may be alternative groups that can "cover the 1's" on a Karnaugh map. Normally the largest groups would be chosen, but sometimes there are alternative equivalently sized groups.

Selecting the best cover

Groups on a Karnaugh map (including single 1's) are called *implicants*. Groups not including groups contained within larger groups are called *prime implicants*. Prime implicants cannot be combined with others to make them bigger. Our problem is to select the prime implicants that best cover the function. Those prime implicants covering at least one 1 that cannot be covered by another prime implicant are necessary for the solution and are called *essential prime implicants*. Those prime implicants which are not essential but nevertheless cover 1's that are not covered by the essential prime implicants are called *non-essential prime implicants* and may be required in the solution. There may also be other prime implicants which cover only 1's already covered by one or more essential prime implicants; these prime implicants are called *(essentially) redundant prime implicants*. These prime implicants are not required in the solution in any circumstances.

A general minimization strategy is first to identify the essential prime implicants. These prime implicants must be in the solution, and can be found by identifying at least one 1 which cannot be covered in any other way. Then it is necessary to select the best set of non-essential prime implicants that will cover the remaining uncovered 1's.

EXAMPLE

Suppose the function:

$$f(A,B,C,D) = \bar{A}\bar{B}\bar{C}\bar{D}+\bar{A}\bar{B}\bar{C}D+\bar{A}\bar{B}CD+\bar{A}B\bar{C}D+\bar{A}BCD+A\bar{B}\bar{C}\bar{D}+A\bar{B}C\bar{D}$$

is to be minimized. Figure 3.20 shows the mapping. Here some 1's on the map can be covered by different groups, while some can only be covered by one group. There are two essential prime implicants, $A\bar{D}$ and $A\bar{B}\bar{D}$, due to the 1's

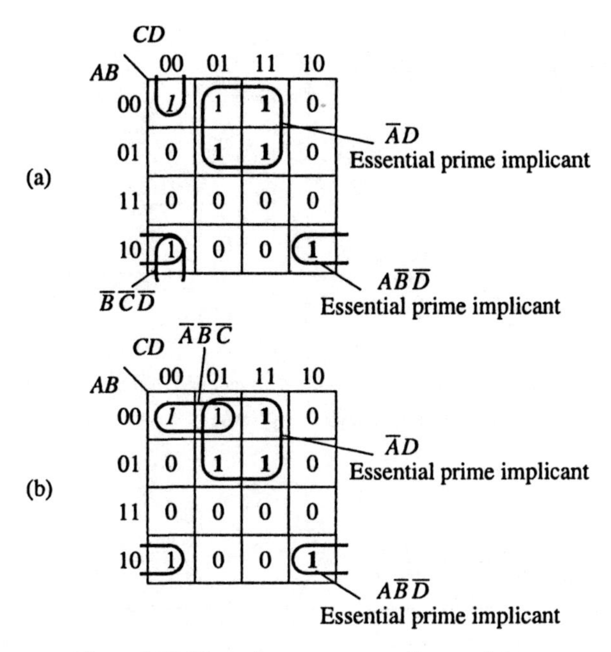

Figure 3.20 Alternative groups on a Karnaugh map

in bold. Hence six 1's are covered, leaving only one 1, shown in italics, to cover. This 1 can be covered by one of two different prime implicants, $\bar{B}\bar{C}\bar{D}$ or $\bar{A}\bar{B}\bar{C}$, so there are two solutions, either:

$$f(A,B,C,D) = \bar{A}D + A\bar{B}\bar{D} + \bar{B}\bar{C}\bar{D}$$

or:

$$f(A,B,C,D) = \bar{A}D + A\bar{B}\bar{D} + \bar{A}\bar{B}\bar{C}$$

Truth table description and the Karnaugh map

There is a direct relationship between the squares on a Karnaugh map and the truth table description of the function; both have one entry for each minterm. Therefore it is a simple matter to map a function onto a Karnaugh map given a truth table description of the function. Figure 3.21 shows the correspondence of truth table entries and squares on the Karnaugh map for a three-variable function, $f(A, B, C) = \bar{A}\bar{B}\bar{C} + \bar{A}\bar{B}C + ABC$.

Minterm numbering in Karnaugh map

Sometimes it is convenient to recognize the minterm numbers of each square of a Karnaugh map. These numbers are shown in Figure 3.22 for three-variable and four-variable maps, assuming variable ordering as given.

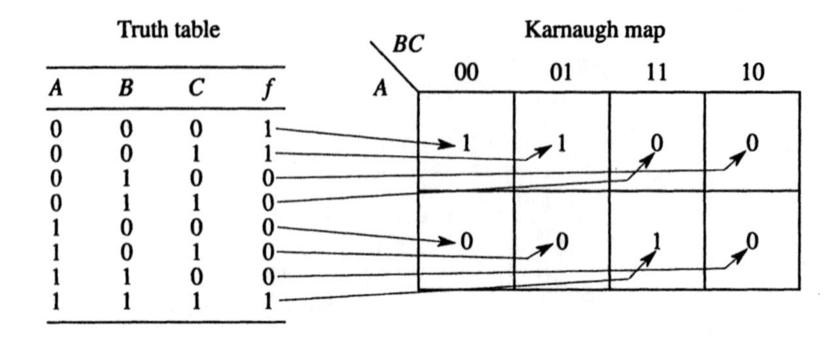

Figure 3.21 Correspondence between a truth table description and a Karnaugh map description of the function $f(A, B, C) = \bar{A}\bar{B}\bar{C} + \bar{A}\bar{B}C + ABC$

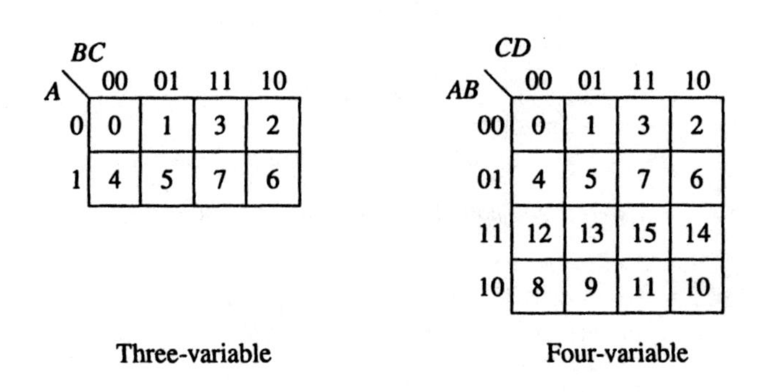

Figure 3.22 Minterm numbering on Karnaugh maps

Functions not in canonical form

So far, functions used in Karnaugh map minimization are canonical sum-of-product expressions; each product term in the expression will map onto one square on the Karnaugh map. A non-canonical expression could be expanded into a canonical expression and then the terms mapped onto the Karnaugh map. However, we can recognize the squares that the expanded terms would cover without actually expanding the terms. For example, in a four-variable function, $f(A, B, C, D)$, the term $\bar{A}\bar{B}C$ corresponds to two adjacent squares on the Karnaugh map, $\bar{A}\bar{B}CD$ and $\bar{A}\bar{B}C\bar{D}$ (because $\bar{A}\bar{B}CD + \bar{A}\bar{B}C\bar{D} = \bar{A}\bar{B}C(D + \bar{D}) = \bar{A}\bar{B}C$). The adjacent squares can be recognized by their Karnaugh map labels (especially with the parentheses labelling in Figure 3.18). With two adjacent squares, one variable will be missing from the corresponding term, the variable which is different in each square. Similarly, four adjacent squares will eliminate two variables. These four squares might be grouped as a 2×2 square, or as a 4×1 rectangle.

EXAMPLE

The non-canonical function:

$$f(A, B, C, D) = A\bar{B}\bar{C} + A\bar{C}D + \bar{A}\bar{B}C + \bar{A}\bar{B}\bar{C} + A\bar{B}C\bar{D}$$

mapped onto a Karnaugh map is shown in Figure 3.23, leading to the minimized function:

$$f(A, B, C, D) = A\bar{C}D + \bar{B}\bar{D} + \bar{A}\bar{B}$$

Notice the mapping procedure is the exact reverse of the minimization procedure, and the square used for each term can overlap in the same way as in the grouping for minimization. Also notice the use of the four corners to form the group $\bar{B}\bar{D}$.

Incompletely specified functions

Sometimes certain combinations of input values cannot occur. A function would be defined only for those values that do occur. This is an *incompletely specified function.*

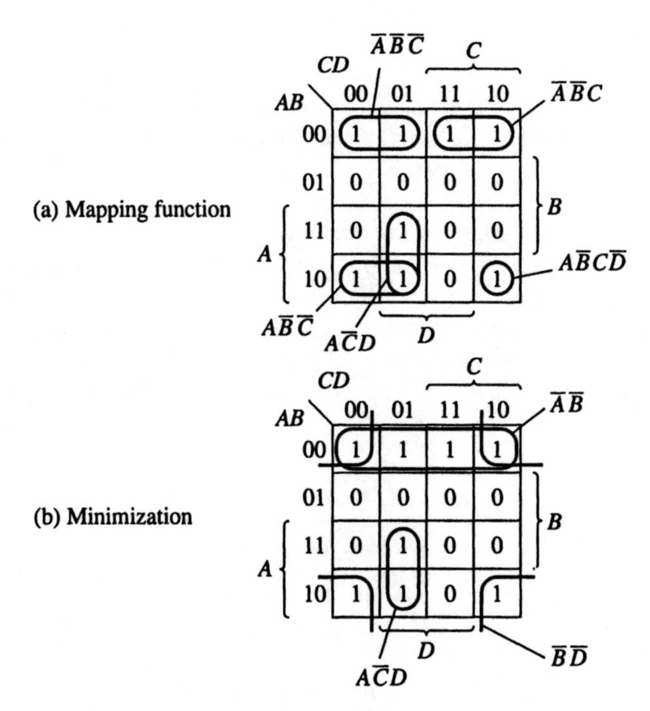

Figure 3.23 Mapping a non-canonical function f(A, B, C, D) = $A\bar{B}\bar{C} + A\bar{C}D + \bar{A}\bar{B}C + \bar{A}\bar{B}\bar{C}$ + $A\bar{B}CD$ onto a Karnaugh map

An incompletely specified function is shown in Table 3.5. In this example, the input combinations 100 and 101 cannot occur and are each marked in the table by an X. These entries are treated as *don't cares*. The same function is described in a Karnaugh map in Figure 3.24. We can take advantage of the don't cares in minimizing the function. Each X can be considered as 0 or 1 whichever leads to the simplest solution. In Figure 3.24, one X is considered as 0 and one is considered as 1.

Table 3.5 *An incompletely specified function*

A	*B*	*C*	*f(A, B, C)*
0	0	0	0
0	0	1	1
0	1	0	1
0	1	1	0
1	0	0	X
1	0	1	X
1	1	0	1
1	1	1	0

A similar situation arises if the input combinations can occur but the output values for these combinations are irrelevant. Again we can use don't cares in defining the output function.

Examples of input combinations not occurring

Unused input combinations can come about when two subsystems are interconnected as shown in Figure 3.25, even if each subsystem is fully specified. The first subsystem accepts logic signals *A*, *B* and *C* and generates logic signals *F* and *G*. Both *F* and *G* are completely specified functions of *A*, *B* and *C*, as described in Table 3.6 but not all combinations of *F* and *G* are generated; $F = 0$, $G = 1$ is not generated in

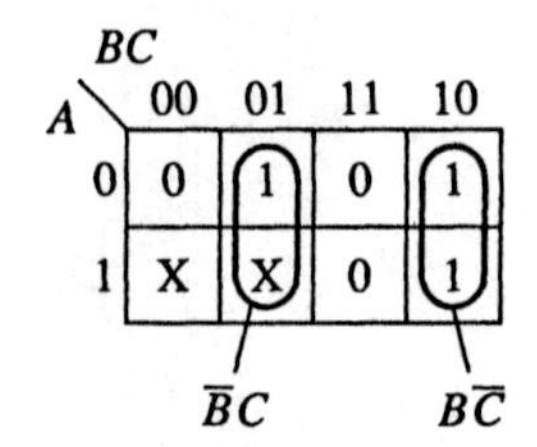

Figure 3.24 Incompletely specified function mapped onto a Karnaugh map

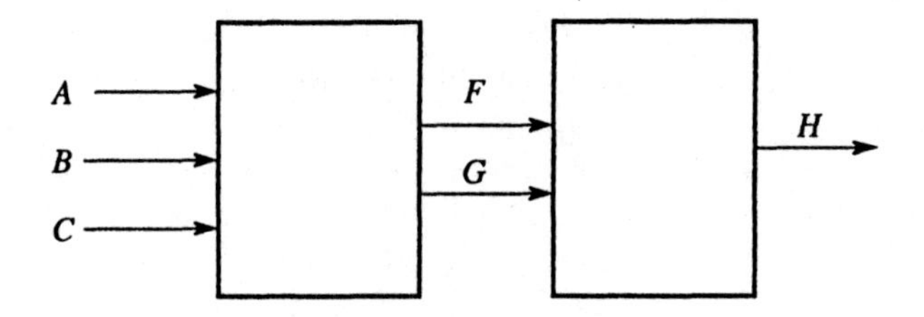

Figure 3.25 Don't cares arising in a logic system

this example, and this pattern becomes a don't care for the design of the second subsystem.

Table 3.6 *Subsystem functions*

A	*B*	*C*	*F*	*G*
0	0	0	0	0
0	0	1	1	0
0	1	0	0	0
0	1	1	1	1
1	0	0	0	0
1	0	1	0	0
1	1	0	1	1
1	1	1	1	1

Another example, widely quoted, is when binary digits are represented in BCD (binary coded decimal, Chapter 1, Section 1.3.3). In BCD, four binary digits are used for each decimal digit, i.e. $0000_2 = 0_{10}$, $0001_2 = 1_{10}$, $0010_2 = 2_{10}$, $0011_2 = 3_{10}$, … $1001_2 = 9_{10}$. The binary patterns 1010_2 through 1111_2 are not used and can be considered as don't cares.

EXAMPLE

Suppose a logic circuit is to be designed which will generate a 1 when the BCD code for eight (1000) or nine (1001) appears. The Karnaugh map of the function is mapped in Figure 3.26, leading to simply the function $f = A$. All invalid BCD codes are mapped as don't cares. Notice that here the minterm numbering of Figure 3.22 is useful.

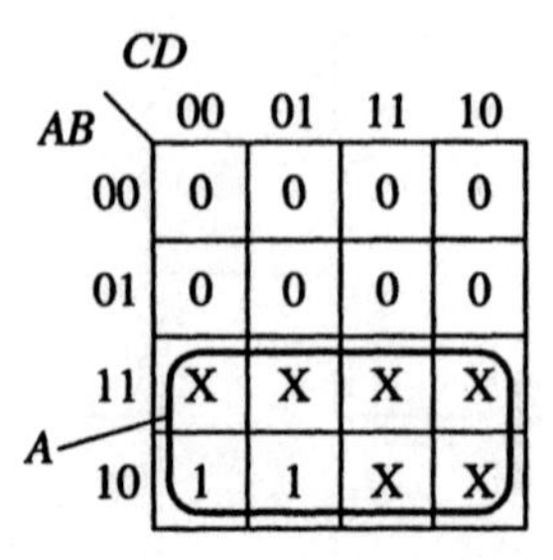

Figure 3.26 Function detecting BCD digits 8 and 9

SELF-ASSESSMENT QUESTION 3.13

What would the function be if only the BCD digit 8 were to be detected?

Minimization using the 0's on the Karnaugh map

The 1's on a Karnaugh map indicate when the function is a 1; similarly the 0's indicate when the function is a 0. Hence the inverse function, $\bar{f}$, can be obtained by simply grouping the 0's rather than the 1's.

EXAMPLE

To minimize the function $f = A\bar{B}C + ABC + AB\bar{C}$, we get the mapping shown in Figure 3.27, the functions:

$f = AB + AC$ from the 1's
$\bar{f} = \bar{A} + \bar{B}\bar{C}$ from the 0's

In this particular case, the inverse function is slightly simpler.

When don't cares exist, they can be considered as 0's or 1's, but if considered differently for the true function, f, and the inverse function, $\bar{f}$, the minimized functions will not be the complement of each other, i.e. $f \neq \bar{\bar{f}}$.

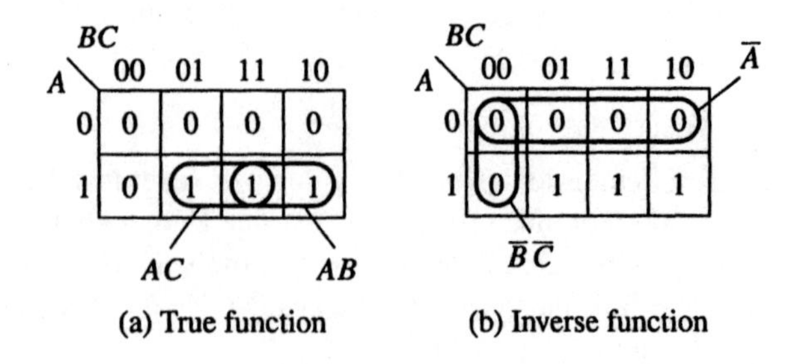

Figure 3.27 Grouping 0's on a Karnaugh map get inverse function

Multi-output minimization

So far, we have considered a single function. It may be that several functions with common variables need to be implemented. In such cases, it may be possible to share gates, i.e. use the output of a gate for more than one function. In order to do this, complete minimization of functions may not be desirable.

EXAMPLE

Suppose there are three functions to implement:

$$f_1 = \bar{A}\bar{B}\bar{C}D + \bar{A}\bar{B}CD + \bar{A}\bar{B}C\bar{D} + \bar{A}B\bar{C}D + \bar{A}BCD + \bar{A}BC\bar{D}$$
$$f_2 = \bar{A}\bar{B}\bar{C}D + \bar{A}\bar{B}C\bar{D} + \bar{A}B\bar{C}D + \bar{A}BC\bar{D} + AB\bar{C}D + ABCD + A\bar{B}\bar{C}D + A\bar{B}CD$$
$$f_3 = \bar{A}\bar{B}\bar{C}D + \bar{A}\bar{B}C\bar{D} + \bar{A}B\bar{C}D + \bar{A}BC\bar{D} + ABCD + A\bar{B}CD$$

The Karnaugh maps for these functions are shown in Figure 3.28. Notice that the fully minimized groups (as shown in dotted lines) are not always used, to facilitate gate sharing.

$$f_1 = \bar{A}D + \bar{A}C\bar{D}$$
$$f_2 = \bar{C}D + ACD + \bar{A}C\bar{D}$$
$$f_3 = \bar{A}\bar{C}D + ACD + \bar{A}C\bar{D}$$

The final logic circuit is shown in Figure 3.29.

SELF-ASSESSMENT QUESTION 3.14

Determine an alternative way of sharing gates in Figure 3.29.

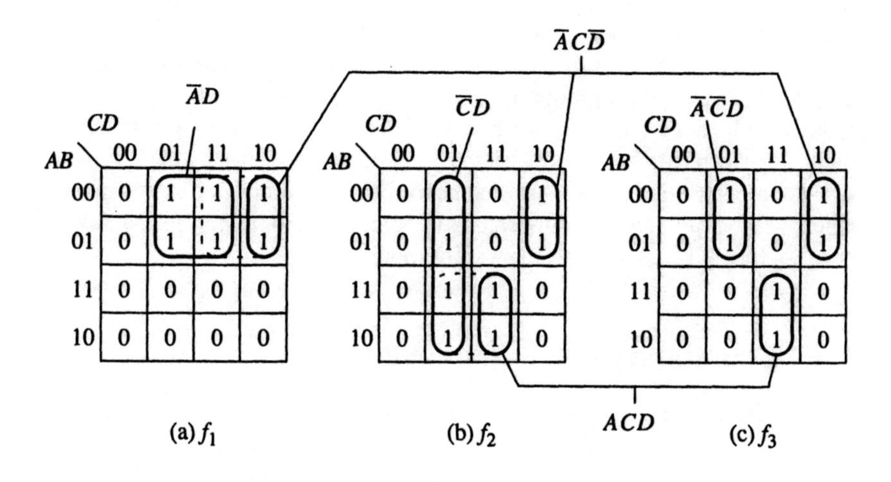

Figure 3.28 Simplifying three functions to reduce number of gates

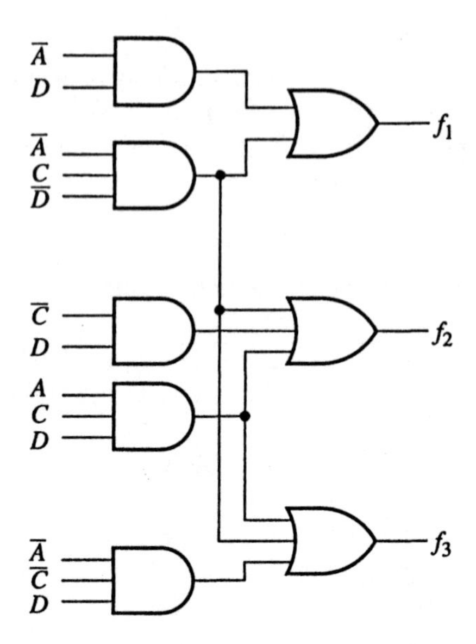

Figure 3.29 Circuit for three functions sharing gates

Large Karnaugh maps

A three-variable map consists of two two-variable maps placed together (side by side), the second two-variable map being a mirror image of the first. Similarly a four-variable map consists of two three-variable maps placed together (one on top the other) with one map being a mirror image of the other. A five-variable map would consist of two four-variable maps placed side by side with one a mirror image of the other as show in Figure 3.30. Hence we can easily construct an n-variable map from two $(n-1)$-variable maps. However, rarely would one use the Karnaugh map method for functions of more than five variables. In fact even a five-variable map is inconvenient because not only can adjacent 1's be combined but also groups from one half of the map with groups in mirror image positions in the other half. For functions with more than four or five variables, we use either the map-entered variable approach (for five or six variables), the Quine-McCluskey method, or more likely, a computer minimization program based on the Quine-McCluskey method or another method suitable for a mechanistic computer program. See the suggested further reading at the end of the chapter for more details.

3.5 MSI combinational logic devices

Some Boolean functions are commonly required and have been manufactured inside a single MSI (medium-scale integration) package. In this section, we shall describe two such common devices, the decoder/demultiplexer and the data selector/multiplexer. Arithmetic circuits, which can be implemented in MSI, will also be examined.

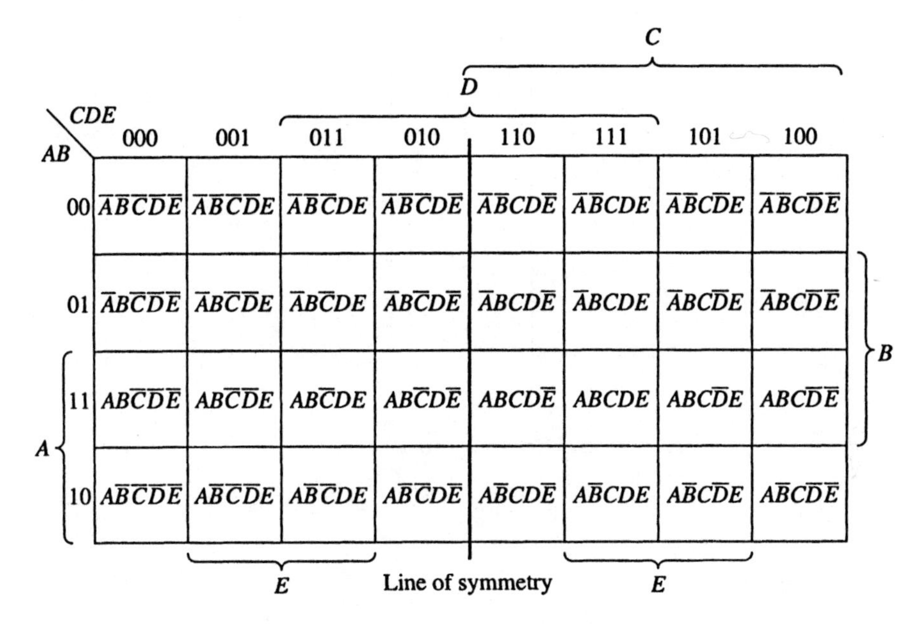

Figure 3.30 Five-variable Karnaugh map

3.5.1 Decoders/demultiplexers

A decoder is designed to recognize the different patterns that can occur on a set of inputs. One output of the decoder is "activated" for each of the possible binary patterns that can occur on the inputs. Suppose there are three inputs. There are eight possible patterns on three inputs and hence eight outputs are needed (as $2^3 = 8$). This decoder would be called a *3-line-to-8-line decoder* and is shown in Figure 3.31.[1] The outputs are commonly *active-low* which means that they are normally at a logic 1 and become a logic 0 to indicate "activation". In the decoder, an output becomes a 0 when the corresponding input pattern has been applied. Active-low outputs are identified here by the use of a bar over the output name. Usually a chip enable input is present, marked $\overline{E}$ in Figure 3.31. The chip "enable" input enables the device and allows outputs to change. The enable input (if *active-low*) would be set to a 0 to enable the device.

A *demultiplexer* is a component similar to a decoder but would be viewed as shown in Figure 3.32. Here there is a single data input line which is routed to one of the outputs dependent upon the pattern on the select inputs. The logic circuits for both a decoder and a demultiplexer are essentially the same. The data input is the enable input in the decoder.

---

[1] There are standard symbols for logic devices such as decoders, see *IEEE Standard Graphic Symbols for Logic Functions* (IEEE, 1984) but these symbols are not universally accepted for logic diagrams, and we shall refrain from using them. Interested readers can find details in the references.

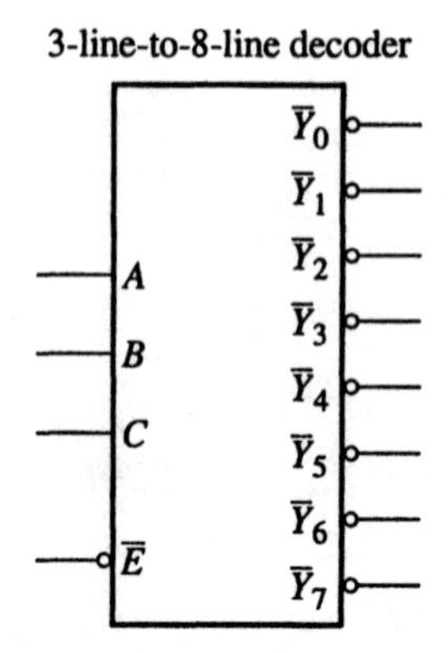

Figure 3.31 3-line-to-8-line decoder

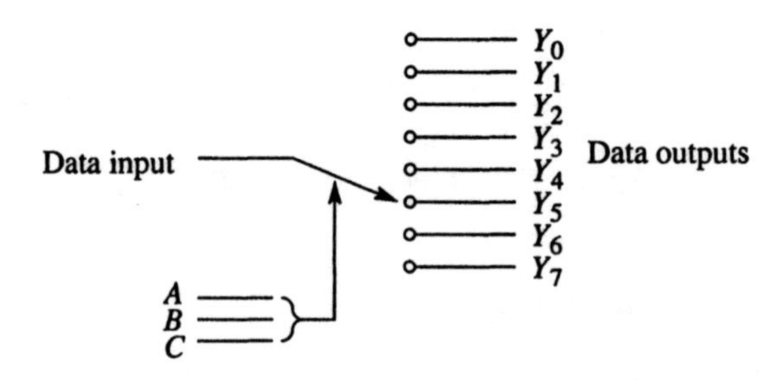

Figure 3.32 Demultiplexer

EXAMPLE

An example is the TTL 74LS138 3-line-to-8-line decoder/demultiplexer. We could describe the function of the decoder/demultiplexer having a single enable/data input in a truth table as shown in Table 3.7. The logic circuit is shown in Figure 3.33. In this circuit, an additional combinational function is available at the data inputs using signals G_1, G_2A and G_2B. The outputs are "inverse" outputs or *active-low* outputs, that is, when selected, the output changes from a 1 (the quiescent state) to a 0, the other outputs remaining a 1. Also note in Table 3.7 the input A is the least significant bit and the input C is the most significant bit if we consider binary numbers being applied to the inputs.

SELF-ASSESSMENT QUESTION 3.15

What function is implemented with G_1, G_2A and G_2B in Figure 3.33?

Table 3.7 *Truth table of a 3-line-to-8-line decoder/multiplexer*

$\overline{E}$	C	B	A	$\overline{Y}_0$	$\overline{Y}_1$	$\overline{Y}_2$	$\overline{Y}_3$	$\overline{Y}_4$	$\overline{Y}_5$	$\overline{Y}_6$	$\overline{Y}_7$
1	X	X	X	1	1	1	1	1	1	1	1
0	0	0	0	0	1	1	1	1	1	1	1
0	0	0	1	1	0	1	1	1	1	1	1
0	0	1	0	1	1	0	1	1	1	1	1
0	0	1	1	1	1	1	0	1	1	1	1
0	1	0	0	1	1	1	1	0	1	1	1
0	1	0	1	1	1	1	1	1	0	1	1
0	1	1	0	1	1	1	1	1	1	0	1
0	1	1	1	1	1	1	1	1	1	1	0

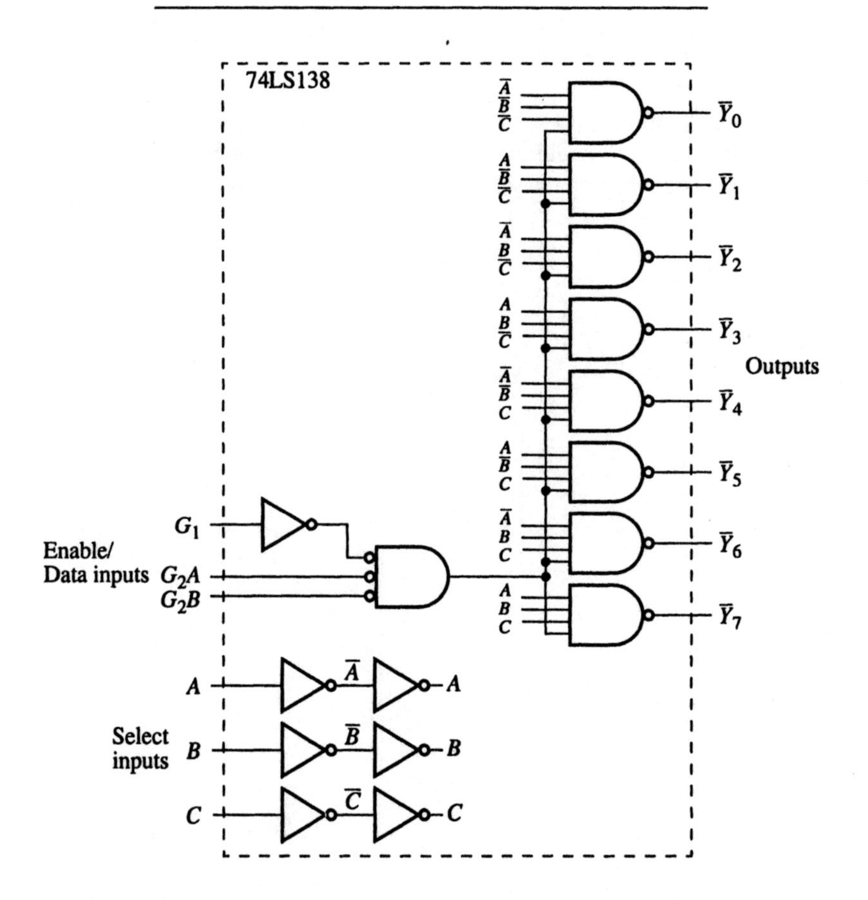

Figure 3.33 3-line-to-8-line decoder/demultiplexer logic diagram

SELF-ASSESSMENT QUESTION 3.16

What logical values must be put on the G_1, G_2A and G_2B enable inputs to enable (activate) the data outputs?

Applications

Address decoder

The primary purpose of a decoder is to "decode", that is, to recognize a particular binary pattern appearing in its inputs and indicate this by activating an output. This action occurs especially in computer interface design where identification "addresses" of memory modules must be recognized. To use a decoder in this application, the signals carrying the memory addresses are applied to the inputs of the decoder and one output is chosen to activate the memory module.

EXAMPLE

Suppose the memory module address to be recognized is 010. Then output $\overline{Y}_2$ would be used as shown in Figure 3.34.

Using a decoder/demultiplexer to implement a combinational logic function

Each output of a decoder/demultiplexer implements one minterm. For example, a 3-line-to-8-line decoder having the select inputs A, B and C (A the least significant bit), and the outputs $\overline{Y}_1, \overline{Y}_2, \overline{Y}_3, \overline{Y}_4, \overline{Y}_5, \overline{Y}_6, \overline{Y}_7$ and $\overline{Y}_8$, implements the functions:

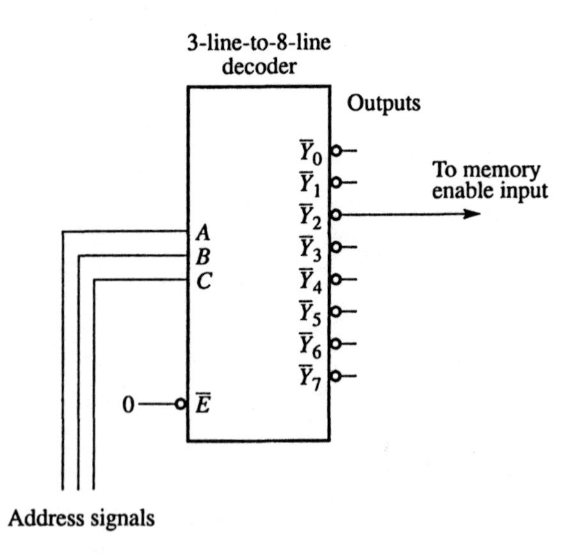

Figure 3.34 Decoding a memory address

$$\begin{aligned}
\overline{Y}_0 &= \overline{\overline{C}\,\overline{B}\,\overline{A}} \\
\overline{Y}_1 &= \overline{\overline{C}\,\overline{B}A} \\
\overline{Y}_2 &= \overline{\overline{C}B\overline{A}} \\
\overline{Y}_3 &= \overline{\overline{C}BA} \\
\overline{Y}_4 &= \overline{C\overline{B}\,\overline{A}} \\
\overline{Y}_5 &= \overline{C\overline{B}A} \\
\overline{Y}_6 &= \overline{CB\overline{A}} \\
\overline{Y}_7 &= \overline{CBA}
\end{aligned}
\qquad \text{or} \qquad
\begin{aligned}
Y_0 &= \overline{C}\,\overline{B}\,\overline{A} \\
Y_1 &= \overline{C}\,\overline{B}A \\
Y_2 &= \overline{C}B\overline{A} \\
Y_3 &= \overline{C}BA \\
Y_4 &= C\overline{B}\,\overline{A} \\
Y_5 &= C\overline{B}A \\
Y_6 &= CB\overline{A} \\
Y_7 &= CBA
\end{aligned}$$

By selecting outputs and using an OR gate, any sum-of-product expression can be formed.

EXAMPLE

Suppose we want to implement the function:

$$f(C, B, A) = \overline{C}BA + C\overline{B}A + CBA$$

Three outputs need to be selected, Y_3, Y_5 and Y_7 (i.e. $\overline{C}BA$, $C\overline{B}A$ and CBA) which connect to the input of a three-input OR gates. Most decoders have active-low outputs so that the circuit becomes as shown in Figure 3.35 using a NAND gate.

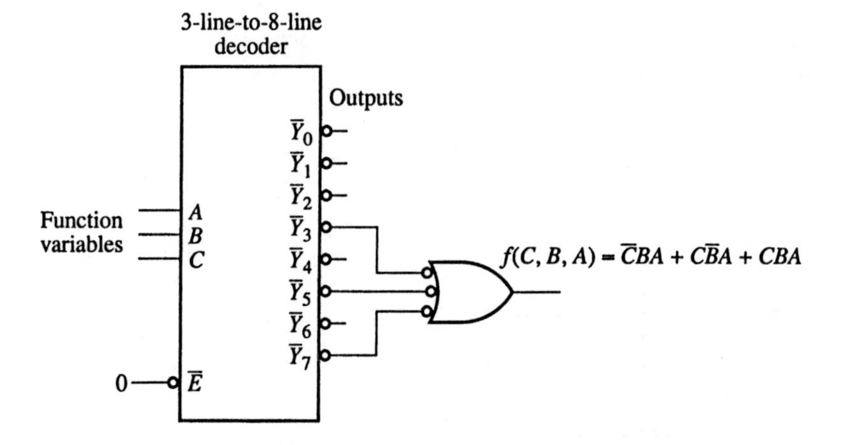

Figure 3.35 Using a decoder to implement a sum-of-product expression

3.5.2 *Multiplexers*

A multiplexer is a logic circuit which allows one of a set of data inputs to be selected and fed into a single output, functionally as shown in Figure 3.36. There are eight data inputs in this example, D_0, D_1, D_2, D_3, D_4, D_5, D_6 and D_7. The single output is *Y*. Additional inputs, *A*, *B* and *C* select one of the eight inputs to connect to the *Y* output. If $CBA = 000$, D_0 would be selected and $Y = D_0$. If $CBA = 001$, D_1 would be selected and $Y = D_1$, and so on (again assuming *A* is the least significant input). A multiplexer having eight data inputs and one data output such as this example is called an *8-line-to-1-line data selector or multiplexer.*

EXAMPLES

A TTL 8-line-to-1-line data selector is the 74LS151 packaged in a 16-pin dual in-line package (providing both true and inverse outputs, *Y* and $\overline{Y}$). The logic diagram of this device is shown in Figure 3.37. Other data selectors in the TTL family include the 74LS153 dual (two in a package) 4-line-to-1-line data selector and the 74LS157 quad (four in a package) 2-line-to-1-line data selector.

Applications

Address decoder

A data selector/multiplexer can be used as for an address decoder, as shown in Figure 3.38. Here a block diagram is used for the data selector with bubbles to show inversion. Commonly the inverse signal is needed to activate the memory module, as shown. In this example, the binary pattern being recognized is 010 on three address bits connected to *A*, *B* and *C* of the decoder. One of the data inputs is set permanently to a logic 1, the data line which corresponding to the required *ABC* pattern. The other data inputs are set permanently to a logic 0. (For TTL, the inputs could be connected to 0 volts for a logic 0 and to 5 volts for a logic 1.) Figure 3.38 shows switches which would allow different addresses to be selected. The connection may also be "hard-wired" links to a logic 0 or a logic 1.

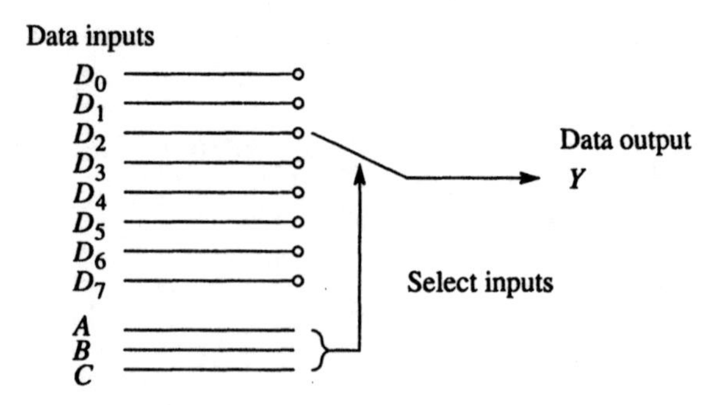

Figure 3.36 Model of a multiplexer

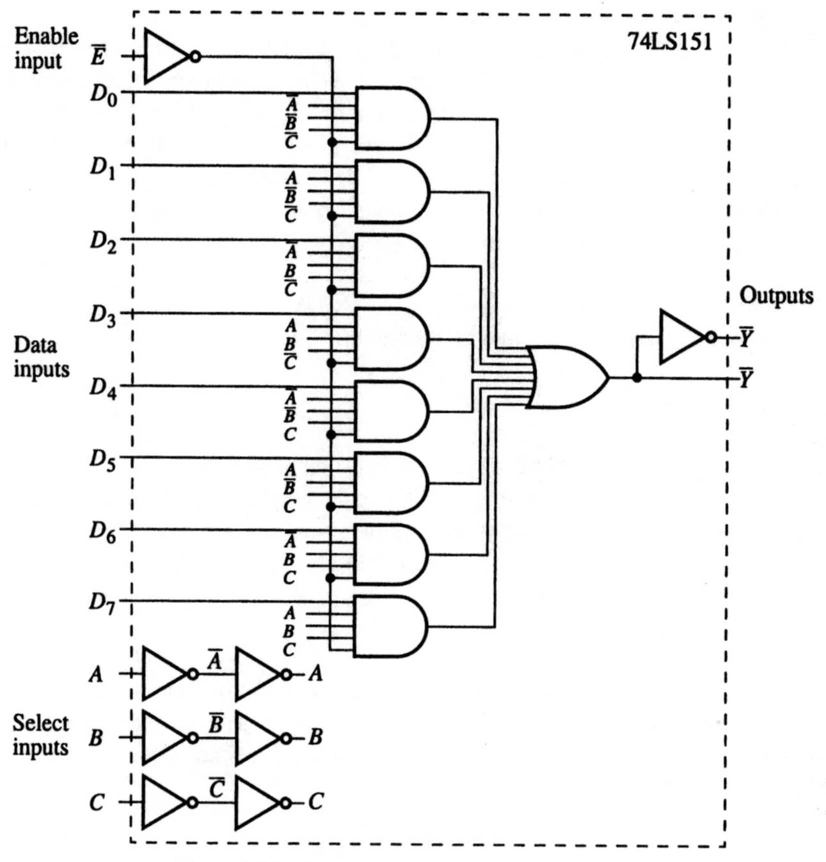

Figure 3.37 8-line-to-1-line data selector/multiplexer

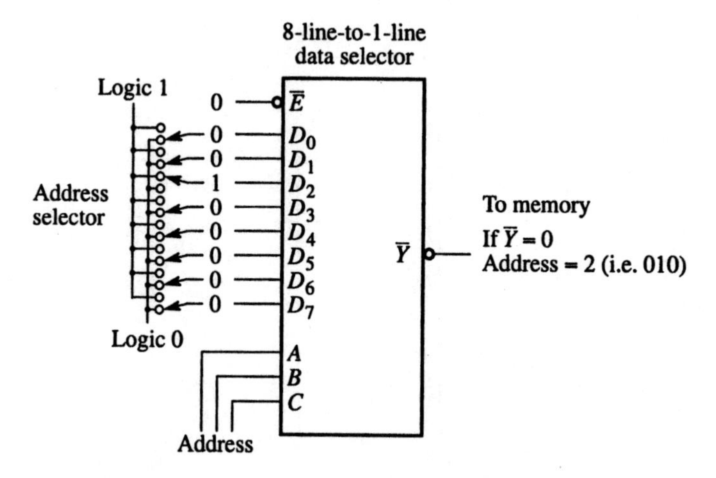

Figure 3.38 Memory address decoder using a data selector

Using a data selector to implement a combinational logic function

Apart from its principal function for selecting one data input to feed to a single output, the data selector can be used to implement a sum-of-product expression. If a data input is permanently set to a 1, and the data input is selected by the appropriate combination of select input values, the Y output will be a 1. Hence by permanently connecting each data input to either a 0 or a 1, depending upon whether the associated product term formed from the select inputs is part of the required function, the Y output will implement the function.

EXAMPLE

Suppose we want to implement the function:

$$f(C, B, A) = C\bar{B}A + \bar{C}\bar{B}A + CBA$$

An 8-line-to-1-line data selector is needed such as shown in Figure 3.37 since there are three variables A, B and C. Three data inputs, D_1 (for $\bar{C}\bar{B}A$), D_5 (for $C\bar{B}A$) and D_7 (for CBA) are set to a 1, the remaining inputs are set to a 0 permanently. The logic diagram is shown in Figure 3.39.

3.5.3 Arithmetic circuits

Computers obviously require logic circuits to perform addition (and other arithmetic operations), and sometimes other digital systems require arithmetic circuits. In this section, let us outline binary addition circuits. Such circuits are often implemented in MSI.

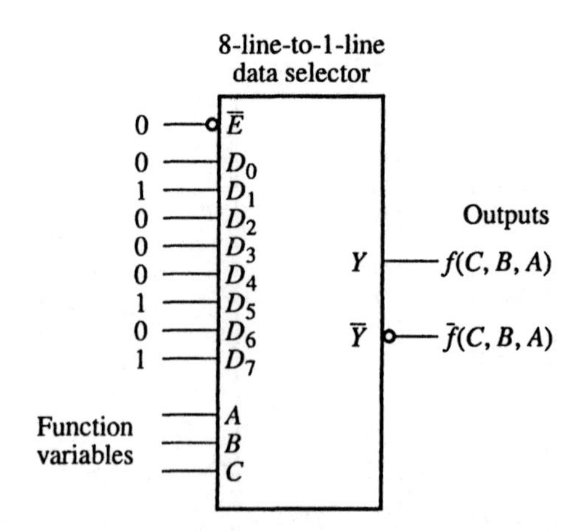

Figure 3.39 Generating a sum-of-product expression using a data selector

Addition

One implementation of binary addition uses the paper-and-paper method described in Chapter 1, Section 1.3.1. First, a circuit is needed that can add together two binary digits, for the least significant bits of the two numbers being added together. This circuit is called a *half adder*. The half adder has two inputs and two outputs and follows the function described in Table 3.8.

Table 3.8 *Half adder truth table*

A	*B*	*SUM*	*CARRY*
0	0	0	0
0	1	1	0
1	0	1	0
1	1	0	1

Second, a circuit is needed which can add together two binary digits with a carry from a previous addition. This circuit is called a *full adder*. The full adder has three inputs and two outputs and follows the function described in Table 3.9.

Table 3.9 *Full adder truth table*

A	*B*	C_{in}	*SUM*	C_{out}
0	0	0	0	0
0	0	1	1	0
0	1	0	1	0
0	1	1	0	1
1	0	0	1	0
1	0	1	0	1
1	1	0	0	1
1	1	1	1	1

If we were to design an addition circuit to add together, say, two n-bit numbers, we would need one half adder and $n - 1$ full adder circuits arranged as shown in Figure 3.40. This arrangement is called a *parallel adder* because pair of digits are added together, i.e. "in parallel" (not strictly, since a carry has to be generated from one stage to the next but the term parallel is always used here). There are other, faster, ways of arranging circuits for adding numbers but most are based upon this arrangement.

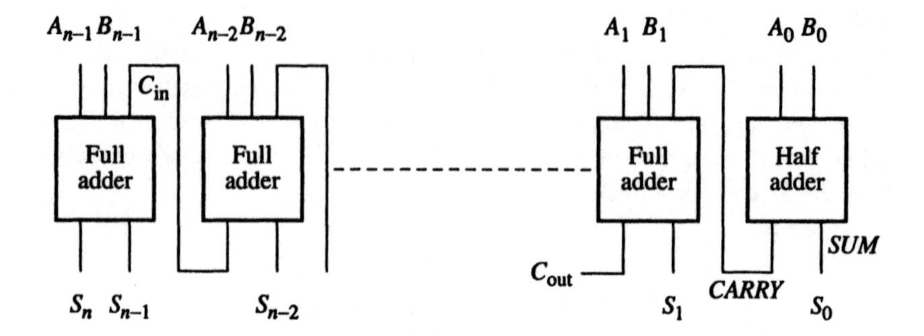

Figure 3.40 Circuit arrangement to add two binary numbers

The logic circuits for the half and full adders can be obtained by mapping the truth tables onto Karnaugh maps and minimizing the functions. It turns out that the half adder functions do not minimize and are:

$$SUM = \bar{A}B + A\bar{B}$$
$$CARRY = AB$$

(in fact just corresponding to the 1's in the output column of the truth tables). The logic circuit for the half adder is shown in Figure 3.41. The sum function is simply the exclusive-OR function.

The full adder functions from the truth tables are:

$$SUM = \bar{A}\,\bar{B}C_{in} + \bar{A}B\bar{C}_{in} + A\bar{B}\,\bar{C}_{in} + ABC_{in}$$
$$C_{out} = \bar{A}BC_{in} + A\bar{B}C_{in} + AB\bar{C}_{in} + ABC_{in} = AB + AC_{in} + BC_{in}$$

Only the C_{out} function minimizes as shown here. The logic circuit for the full adder is shown in Figure 3.42. It is possible to construct a full adder from two half adder circuits (see Wilkinson, 1992).

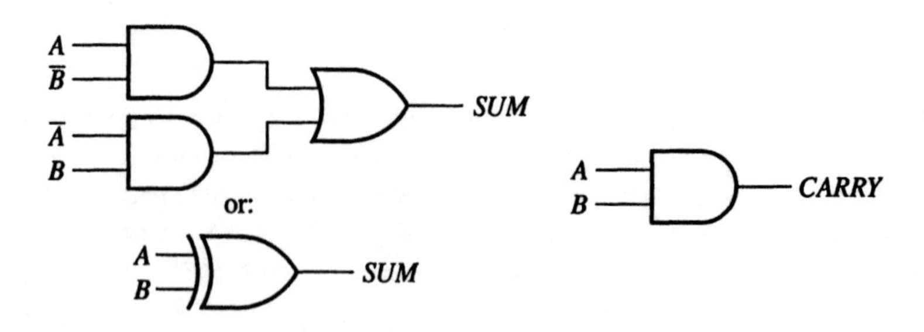

Figure 3.41 Half adder

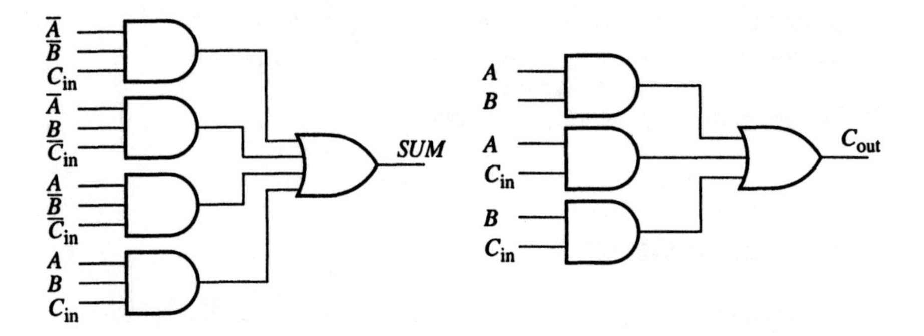

Figure 3.42 Full adder

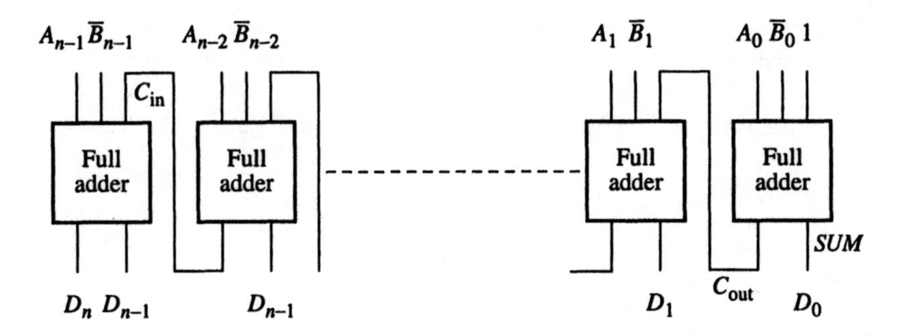

Figure 3.43 Circuit arrangement to subtract two binary numbers

Subtraction

Subtracting the number B from the number A is done with 2's complement arithmetic by adding the negative of B to A. The negative representation can be achieved by inverting the digits of B and adding 1. Adding 1 can be done very efficiently in the parallel adder by changing the first stage from a half adder to a full adder and setting the carry input to this first stage to 1. Hence subtraction can be performed with a parallel adder as shown in Figure 3.43.

3.6 Summary

Covered in this chapter are:

- Different ways to implement Boolean sum-of-product and product-of-sum expressions.

- Simplification of Boolean expressions using algebraic methods.
- Simplification of Boolean expressions using the Karnaugh map method.
- MSI multiplexers and demultiplexers
- Simple addition and subtraction circuits.

3.7 Tutorial questions

3.1 The output of a logic circuit is a 1 only when any one of the following patterns is present at its three inputs:

000, 001, 010, 011, 110

Draw the truth table of the output function. Obtain the minimal Boolean expression, and develop a logic circuit to generate the output signal.

3.2 What Boolean function does the circuit shown in Figure 3.44 generate?

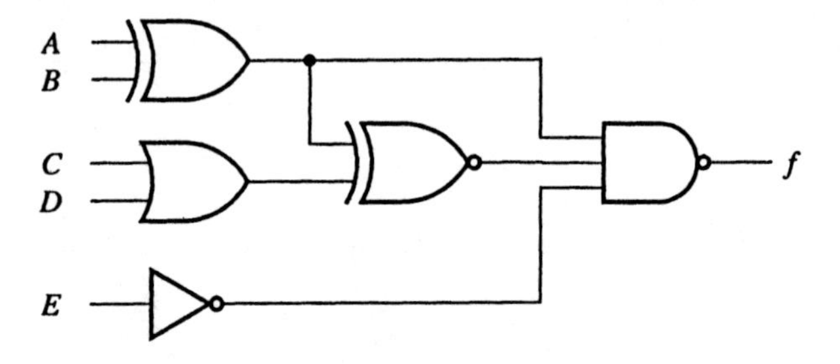

Figure 3.44 Circuit for Question 3.2

3.3 Expand the following expressions into canonical form:

(a) $AB + \bar{A}\bar{B} + C$
(b) $(A + B)(\bar{A} + \bar{C})C$

3.4 Convert the following canonical sum-of-product expressions into canonical product-of-sum expressions:

(a) $f(A, B, C) = AB\bar{C} + A\bar{B}\bar{C} + \bar{A}B\bar{C} + ABC$
(b) $f(A, B, C, D) = \Sigma(0, 3, 5, 9)$

3.5 Implement the following functions using NAND gates only.

(a) $f(A, B, C) = \overline{A}\,\overline{B}\,\overline{C} + A\overline{B}\,\overline{C} + AB\overline{C} + ABC$
(b) $f(A, B, C) = (\overline{A} + \overline{B} + \overline{C})(A + \overline{B} + \overline{C})(A + B + C)$

3.6 Repeat Question 3.5 using NOR gates only.

3.7 Draw a two-level logic circuit for each of the following functions, assuming that the complements of variables are available:

(a) $f(A, B, C) = \overline{A}\,\overline{B}(\overline{C} + ABC) + \overline{AB + C}$
(b) $f(A, B, C) = (\overline{A} + \overline{B} + \overline{C})(A + \overline{B} + \overline{C})(\overline{\overline{A} + B + \overline{C}})$

3.8 Implement the following expressions using 2-input NAND gates only:

(a) $f(A, B, C, D) = A\overline{B}C\overline{D} + A\overline{B}\,\overline{C}D + A\overline{B}\,\overline{C}\,\overline{D}$
(b) $f(A, B, C, D) = (A+B+C+\overline{D})(A+B+\overline{C}+D)(A+\overline{B}+C+D)(\overline{A}+B+C+D)$

3.9 Figure 3.45 uses alternative logic symbols. What is the Boolean output function, F, of the circuit?

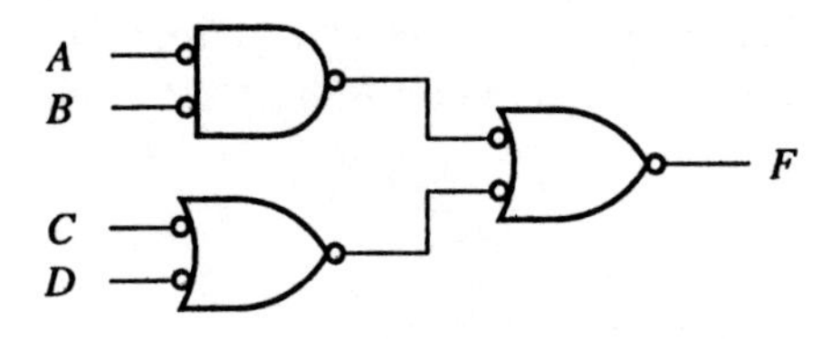

Figure 3.45 Circuit for Question 3.9

3.10 Minimize the following expressions algebraically:

(a) $f(A, B, C) = (A + BC)(B + C)(ABC + BC)$

(b) $f(X, Y, Z) = \overline{(\overline{X + YZ})(XY(\overline{\overline{\overline{Z}} + YZ}))}$

3.11 Prove the Consensus theorem algebraically.

3.12 Use only the Consensus theorem to simplify the following:

(a) $A\overline{B}\,\overline{C} + A\overline{B}D + \overline{B}CD + \overline{A}CD + A\overline{C}\,\overline{D}$
(b) $(A + \overline{B})(A + C)(A + D)(B + C)(B + D)(C + \overline{D})$

3.13 Find a simpler circuit for the logic circuit shown in Figure 3.46 that creates the same output.

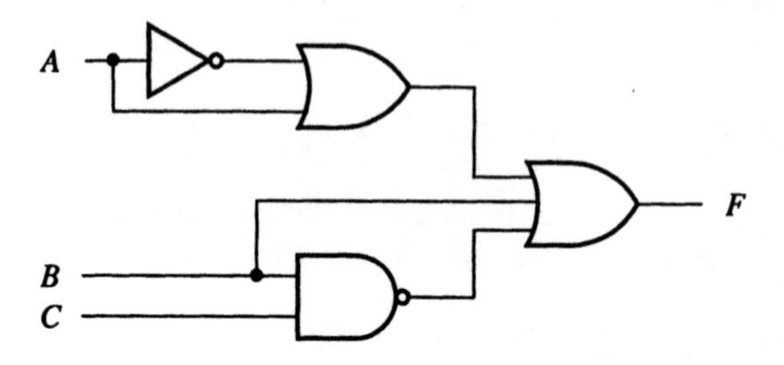

Figure 3.46 Circuit for Question 3.13

3.14 A logic circuit accepts two 2-bit numbers and has one output, Z, which is a logic 1 only when the two 2-bit numbers applied to the circuit are equal. Design a two-level circuit using AND gates and an OR gate, assuming that the complements of the inputs are available.

3.15 Design a two-level logic circuit which adds together four 1-bit digits using AND gates and an OR gate. (Start with a table listing the inputs and the required outputs.)

3.8 Suggested further reading

IEEE, *Standard Graphic Symbols for Logic Functions*, Std. 94–1984, New York: IEEE, 1984.

Roth, C. H., *Fundamentals of Logic Design, 3rd Edition*, West Publishing Company: St. Paul Minnesota, 1985.

Wilkinson, B., *Digital System Design, 2nd edition*, Prentice Hall: London, 1992.

CHAPTER 4

Flip-flops and counters

Aims and objectives

In this chapter, we introduce logic components whose outputs depend not only on the present logic input values but also on previous logic input and output values. These components can remember the past events and have a response dependent upon past events, an essential feature for designing complex logic systems.

4.1 Sequential circuits

So far we have considered logic circuits whose output values depend upon the values on the inputs. For example, the output of a basic AND gate will be a 1 immediately two 1's are applied to the inputs. We have called such circuits *combinational circuits* because the output value(s) will depend upon certain combinations of input values. Often, however, we need a logic circuit whose output values depend upon previous values of the outputs (and hence previous values of the input). A common example is a binary counter whose outputs follow a binary sequence, such as 000, 001, 010, 011, 100, 101, ... The output pattern after 000 is 001. The output pattern after 001 is 010, and so on. The change from one output pattern to the next is initiated by a *clock signal* applied to the counter. The clock signal changes from 0 to 1 and from 1 to 0 at regular intervals. The change from one number to the next number will occur when the clock changes from a 0 to a 1 or from a 1 to a 0 depending upon the design of the components. Figure 4.1 shows the binary-up sequence if the changes of output occur at the time the clock changes from a 0 to a 1. The outputs are given by $Q_0Q_1Q_2$.

SELF-ASSESSMENT QUESTION 4.1

Draw Figure 4.1 if the changes of output occur at the time the clock changes from a 1 to a 0.

The particular sequence in Figure 4.1, a binary increasing sequence, has three digits requiring three outputs, one for each bit. Whether an output bit is a 0 or a 1 will depend upon the previous values of all three outputs. The least significant bit will become a 1 after the number 000 or 010 or 100 or 110, and will become a 0 after the

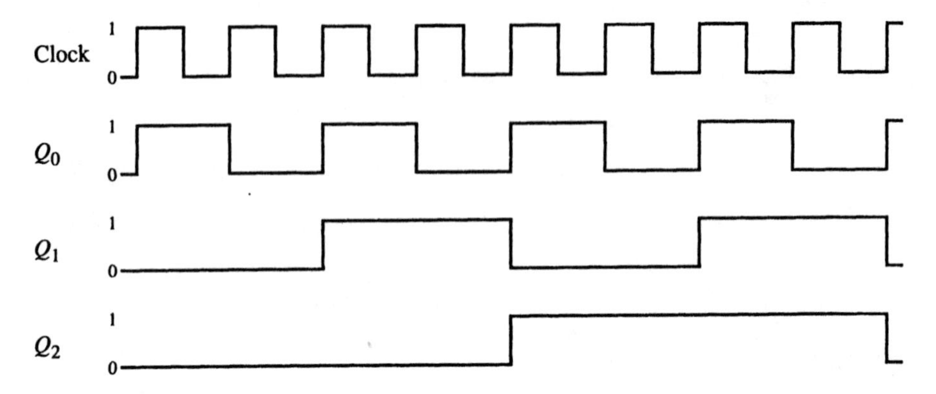

Figure 4.1 Binary-up counting sequence

numbers 001 or 011 or 101 or 111.

Counters are in the classification called *sequential logic circuits*. Sequential logic circuits are characterized by output values which depend not only on present input/output values but also on past input/output values. A sequential logic circuit whose output changes are initiated by an input clock signal and in which the outputs change immediately upon the required clock signal transition is called a *synchronous sequential logic circuit*. In this chapter, we shall examine the design of synchronous sequential logic circuits. Synchronous sequential logic circuits form the majority of sequential systems because having a clock signal to synchronize outputs greatly simplifies the design. Computers, for the most part, are synchronous sequential systems. However, there are some instances where using a clock to synchronize operations is not appropriate or possible. A sequential logic circuit that does not use a synchronizing clock signal is called an *asynchronous sequential circuit*. A treatment of asynchronous logic circuit design can be found in texts such as Wakerly (1990) or Wilkinson (1992). We shall limit our discussion to synchronous sequential logic circuits, but should mention in passing that very basic synchronous sequential logic circuits are in fact asynchronous sequential logic circuits internally, the clock input being simply another asynchronous input.

A sequential logic circuit must have some form of memory to remember past values. Let us first look how a memory cell can be designed, and then how memory can be incorporated into the basic building blocks for sequential circuits.

4.2 Memory design using gates

A logic circuit can maintain a constant output value by the use of feedback whereby the output is connected to the inputs in such a way to reinforce the output value. For example, suppose an AND gate has 1's on its inputs. This produces a 1 on the output.

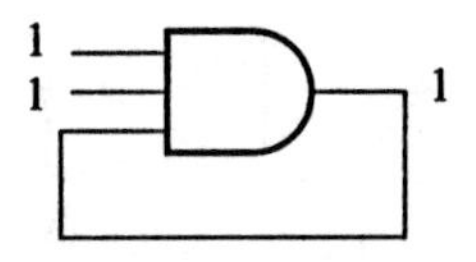

Figure 4.2 Potential memory design

If this 1 is fed back to create a 1 on an input with the other inputs kept at 1, the output will maintain a 1 output, as shown in Figure 4.2. If the output was initially a 0, a 0 would be fed back and reinforce and maintain the 0 output. Whether the output is a 0 or a 1 is indeterminate, that is, it could be a 0 or a 1 when the circuit is first switched on depending upon internal design of the gates, and the output value would remain unchanged if the other AND gate inputs are a 1.

If one of the other inputs of the AND gate were set at a 0, then the output would become a 0 and remain a 0, but we cannot force the output to become a 1 with these inputs only. To be able to force the output to become either a 0 or a 1 requires a circuit such as in Figure 4.3. Here, we have used two NAND gates rather that a single AND gate, and two inputs labelled $\overline{R}$ (reset) and $\overline{S}$ (set). This gives us the ability to force the output to a 0, by applying a 0 the reset input $\overline{R}$ while maintaining $\overline{S}$ at a 1, or to force the output to a 1 by applying a 0 to the set input $\overline{S}$ while maintaining $\overline{R}$ at a 1. Notice that the inputs are labelled with overbars to indicate that a 0 is necessary to cause action. We call such signals *active-low signals* as opposed to *active-high signals* where a 1 causes action. Once the output has been forced to a 0 or a 1, the input ($\overline{S}$ or $\overline{R}$) can return to a 1 and the output will remain unchanged

Often it is convenient to have both a true output, Q, and a complementary $\overline{Q}$ output which in normal operation of the memory circuit has the opposite logical value to Q, i.e. if $Q = 0$, $\overline{Q} = 1$ and if $Q = 1$, $\overline{Q} = 0$. The complementary output can be obtained in Figure 4.3 from the output of the first NAND gate. The circuit configuration is usually drawn as shown in Figure 4.4. It is also sometimes convenient to have active high input signals which can be achieved with the use of additional NAND gates (or NOT gates) as shown in Figure 4.5. Memory designs such as in Figure 4.3, Figure 4.4 and Figure 4.5 are called *latches* and appear in more complex memory designs called *flip-flops*.

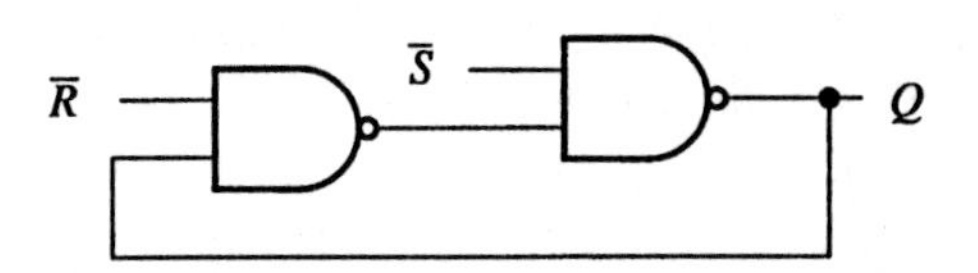

Figure 4.3 Set–reset memory design

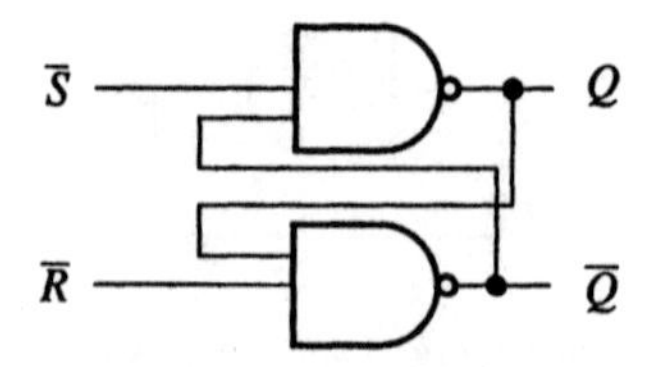

Figure 4.4 Memory design with true and complementary outputs

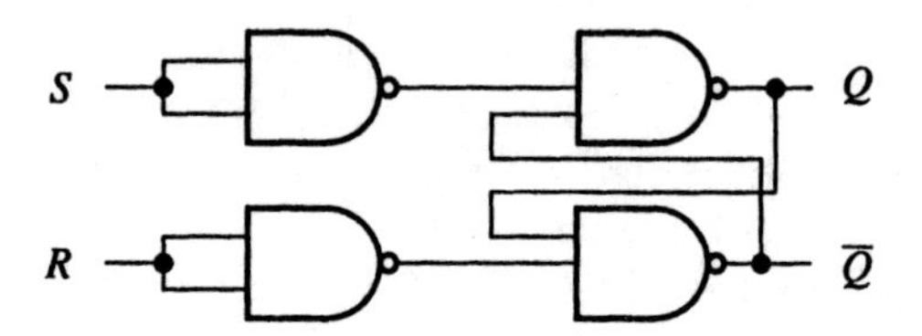

Figure 4.5 Memory design with true and complementary outputs and active high inputs

SELF-ASSESSMENT QUESTION 4.2

What happens if $R = 0$ and $S = 0$ in Figure 4.5?

4.3 Flip-flops

The latch design in the previous section will store one binary value, but has the disadvantage that the outputs will change immediately one of the inputs changes to a 0. In a synchronous sequential circuit, we want the output changes to be synchronized with a clock signal. Such memory designs are usually called *flip-flops*. A flip-flop stores a single bit by producing an output of a 0 or a 1 continuously until changed by conditions on the inputs and a clock signal transition. The term flip-flop comes from the operation of a flip-flop – the output flips from a 0 to a 1 or flops from 1 to a 0. As with the latches previously, the flip-flop output is usually given the letter Q for identification, and a complementary $\overline{Q}$ output is often present which in normal operation of the flip-flop has the opposite logical value to Q.

The flip-flop is the basic building block of sequential circuits. There are several types of flip-flop, just are there are several types of gate. Each type of flip-flop is characterized by a truth table describing the relationship between the inputs and the outputs. Normally only Q is listed (not $\overline{Q}$).

4.3.1 S–R flip-flop

The S–R flip-flop (sometimes called an R–S flip-flop) has essentially the same characteristics of the memory latch in that it has two inputs, named *S* for (set) and *R* for (reset). The *S* input when a 1 will set the output to a 1, while the *R* input when a 1 will reset the output to a 0. Synchronous operation requires an additional clock input and only after a specified clock transition occurs will the outputs take on the required values – before the clock transition occurs, the outputs will not change even if the *S* and *R* inputs change.

Truth table

Flip-flops can be described by a truth table. The truth table of an *S–R* flip-flop is shown in Figure 4.6 where Q_+ indicates the value of Q after the activating clock transition. Q_- is sometimes used to indicate the value of Q before the activating clock transition. X indicates an undefined output.

S	R	Q_+
0	0	$Q_{(-)}$
0	1	0
1	0	1
1	1	X

Figure 4.6 Truth table of S–R flip-flop

Characteristic equation

Flip-flops can also be described by an equation called the *characteristic equation*, which is the relationship between the Q output and inputs. The characteristic equation of an *S–R* flip-flop is:

$$Q_+ = Q\bar{R} + S$$

The logic circuit can be obtained from the characteristic equation.

SELF-ASSESSMENT QUESTION 4.3

Show that the characteristic equation does indeed correspond to the *S–R* flop-flop logic diagram and truth table.

Level triggering

The simplest form of clock activation is to use the 1 level of the clock; when the clock becomes a 1, the outputs assume their values according to *S* and *R*. This is known as *level triggering*. Figure 4.7 shows a level-triggered *S–R* flip design, based upon the memory design of Figure 4.5. When the clock input, CL is at a 0, the outputs of the first row of NAND gates will be at a 1 irrespective of the logic levels on *S* and *R*.

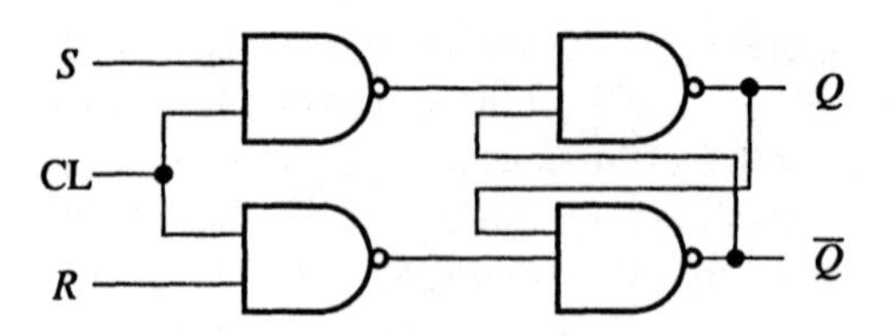

Figure 4.7 Level-triggered S–R flip-flop design

When CL becomes a 1, the values on S and R enter the circuit and the outputs will assume the corresponding input values. During the time that CL = 1, it is assumed that S and R will not change as, if they do, the outputs could be affected. While CL = 0, S and R can change without immediately affecting the outputs, though we would not want to change them just as CL changes to a 1 if the synchronizing nature of the clock is to be maintained. The term ***transparent latch*** is sometimes used for this design.

Sample timing of level triggering is shown in Figure 4.8. Notice how it is necessary for the inputs to remain steady while the clock signal is a 1 (high). Also notice that before the time of transition to a 1, the values of S and R can change.

If we simply want to store the value of one binary digit (which could be a 0 or a 1), then instead of having two inputs, S and R, only one input is needed, a data input, which specifies a 0 or a 1 (in addition to the clock input). Figure 4.9 shows how this might be achieved using the latch design of Figure 4.7. The problem in Self-assessment question 4.1 (given earlier) that occurs when both S and R were at a 1 simultaneously is eliminated.

Level triggering is not particularly convenient because it demands constraints on

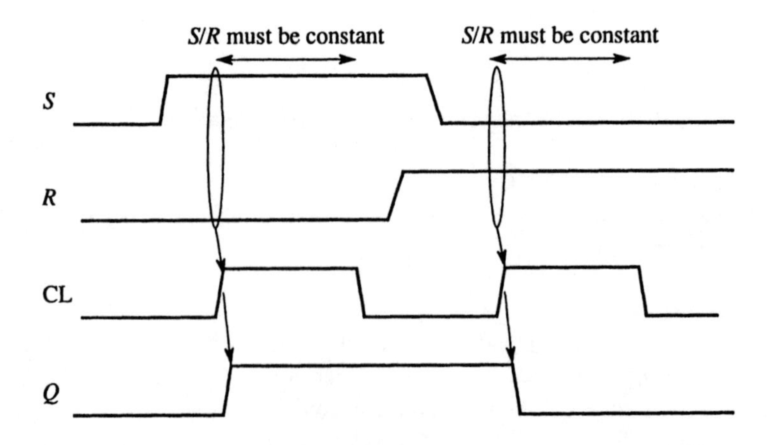

Figure 4.8 Sample timing of S–R flip flop with level triggering

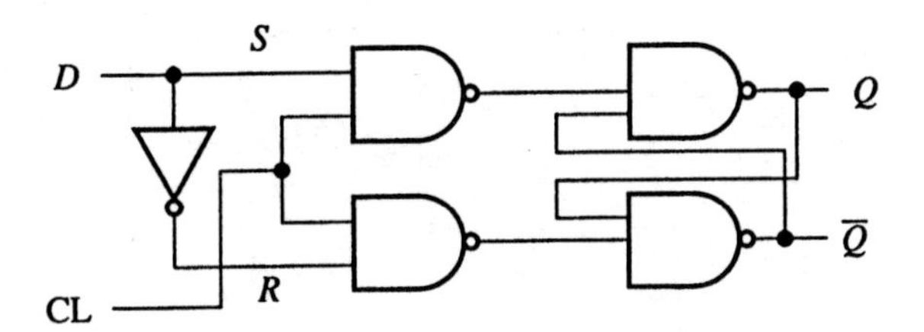

Figure 4.9 Level-triggered D latch design

when the inputs may change. The inputs must not change while the clock is a 1 or at the instant that the clock changes to a 1 or to a 0. More complex triggering mechanisms exist which remove these constraints but with additional cost (more components). Such triggering mechanisms are usually associated with the following two types of flip-flops, *D*-type flip-flops and *J–K* flip-flops, which we shall now describe.

4.3.2 D-type flip-flop

The flip-flop which is directly equivalent to the *D* latch of Figure 4.9 is called a *D-type flip-flop*. The Q output simply becomes the value on the *D* input after the activating clock transition. Figure 4.10 gives the truth table of the flip-flop where Q_+ indicates the value of Q after the activating clock transition. The characteristic equation of the *D*-type flip-flop is simply:

$$Q_+ = D$$

(In this case, the characteristic equation does not immediately suggest a circuit configuration.)

Edge triggering

There needs to be a mechanism whereby the output changes on a transition of the clock signal (so-called *edge triggering*) and the inputs are allowed to change at other times without affecting the output. An early approach to this problem was to use two latches in cascade. The first latch, the *master*, captures the input values when clock

D	Q_+
0	0
1	1

Figure 4.10 Truth table of D-type flip-flop

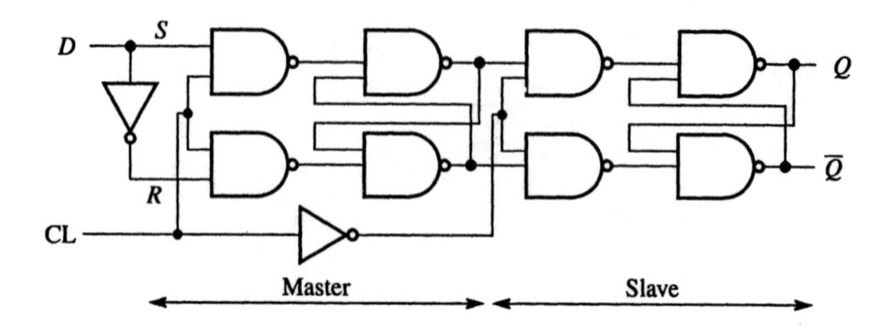

Figure 4.11 Master–slave design (historical)

is a 1, and the second latch, the *slave*, uses the first latch's outputs when the clock is a 0. The outputs will change on the negative transition of the clock. The logic arrangement for a *D*-type flip-flop is shown in Figure 4.11, and the timing in Figure 4.12. This *master–slave* design requires two complete latches and still does not eliminate the constraint that the inputs must not change while the clock is high. Also the output is delayed until the falling edge of the clock signal. More details of this design can be found in the references. (The master–slave design was more common for *J–K* flip-flops, see Section 4.3.3.)

A more convenient and modern form of triggering for flip-flops is the true *edge-triggered* design in which a transition from one specified logic level to the other logic level causes the outputs to change (if they are to change). The outputs are not affected at *any other time* except at the time of the transition and the values on the inputs (just *D* in this case) can change at any other time. The value of input(s) at the time of the clock transition determines the output value. Implementing this is rather complex and the details not particularly important as usually we shall be using packaged flip-

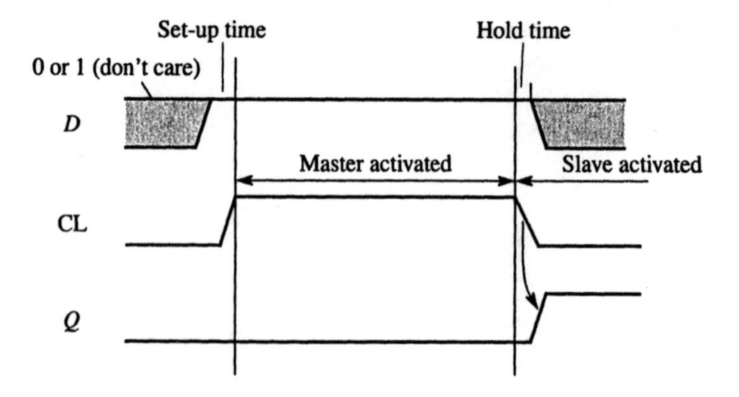

Figure 4.12 Master–slave D-type flip-flop timing

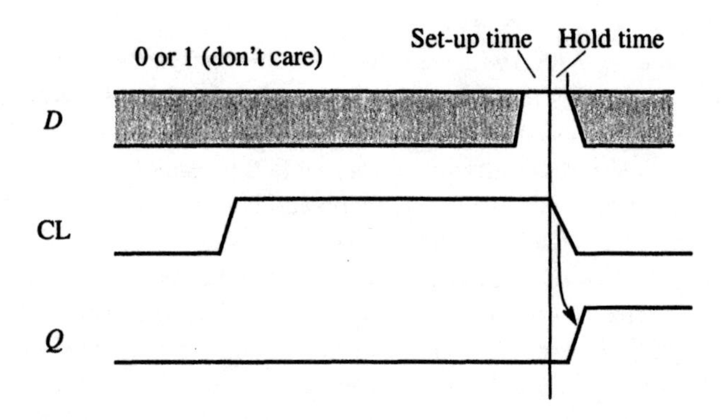

Figure 4.13 Negative edge-triggered D-type flip-flop timing

flops, whereas the latches in the previous section are sometimes constructed using separate gates.

There are two forms of edge triggering. In *positive edge triggering*, the activating transition is from a logic 0 to a logic 1. In *negative edge triggering*, the activating transition is from a logic 1 to a logic 0. Both forms are common. Figure 4.13 shows the timing of a negative edge-triggered D-type flip-flop with $D = 1$. The outputs will only change on the negative transition of the clock, and the change is determined by the value on D at the time of the clock transition. In practice, there will be a very short data *set-up time* period immediately prior to the transition during which the input should not change. Occasionally, there may also be a very short data *hold time* period immediately after the clock transition during which the input should not change, dependent upon the internal design of the flip-flop. Positive edge triggering is similar except that the data set-up and hold periods are around the positive transition of the clock.

Flip-flops are fundamental components of logic systems and logic symbols are used to denote them. Symbols for D-type flip-flops are shown in Figure 4.14. The form of triggering is shown in the symbol, a circle for negative edge triggering. (Master–slave triggering would need a separate notation.) Sometimes the form of triggering is not shown.

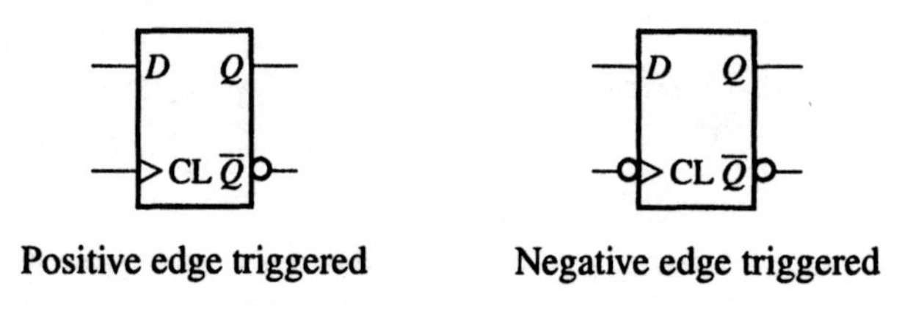

Figure 4.14 D-type flip-flop symbols

Asynchronous set and reset inputs

All flip-flops can be provided with separate (asynchronous) set and reset inputs to set the output to a 1 and reset the output to a 0 immediately and not in synchronization with the clock. This is very useful when the flip-flips are used in counters (see later). Normally the set and reset inputs are active low (a 0 causing output action). The set and reset inputs are marked usually on the top and bottom of the flip-flop symbol as shown in Figure 4.15 (for a *D*-type flip-flop).

Examples of packaged TTL positive edge-triggered *D*-type flip-flops include the 74LS74 dual *D*-type flip-flops, the 74LS175 quad *D*-type flip-flops (four in one package), the 74LS174 hex *D*-type flip-flops (six in one package), and the 74LS374 octal *D*-type flip-flops (eight in one package). In devices where there are more than two flip-flops in one package, all the flip-flops usually have a common clock input.

Flip-flop state diagram

Sequential circuits exist in defined *states*. All practical sequential circuits have a finite number of states, hence the term *finite state machine* for describing practical sequential circuits. A flip-flip can exist in one of two states: one when the output is a 0 and one when the output is a 1. A state change will be initiated by a specified change of inputs and the activating clock transition (for synchronous sequential circuits). We can illustrate the states of a sequential circuit and the conditions for changing from one state to another in a *state diagram.*

A state diagram of a *D*-type flip-flop is shown in Figure 4.16. This particular form of state diagram is named a *Moore model state diagram.* In the Moore model state diagram, each state is shown as a circle with two numbers inside, one shows an arbitrary state number, say 1, 2, etc., and one shows the output(s) of the circuit when in that state. Different states may have the same outputs or different outputs. In flip-flops, there are two states, 1 and 2, each with a unique output. In state 1 the output is a 0 and in state 2 the output is a 1. Transitions from one state to another are shown as arcs between the state circles, labelled with the inputs that are necessary to cause the transition (excluding the clock input which is assumed in synchronous sequential circuits). In our flip-flop, if the current state is state 1 ($Q = 0$) the one input, $D = 1$, will

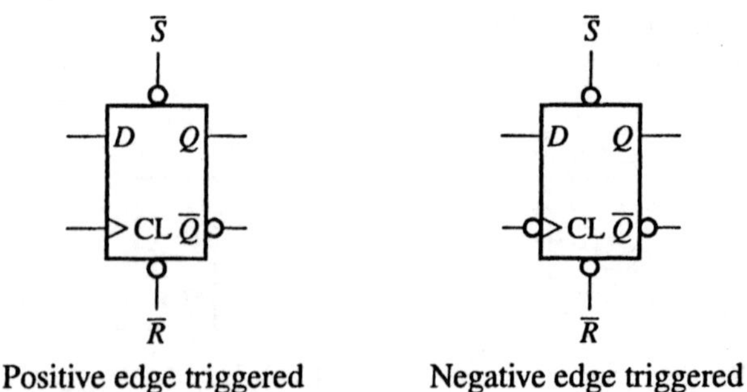

Figure 4.15 Symbols of D-type flip-flops with asynchronous set and reset inputs

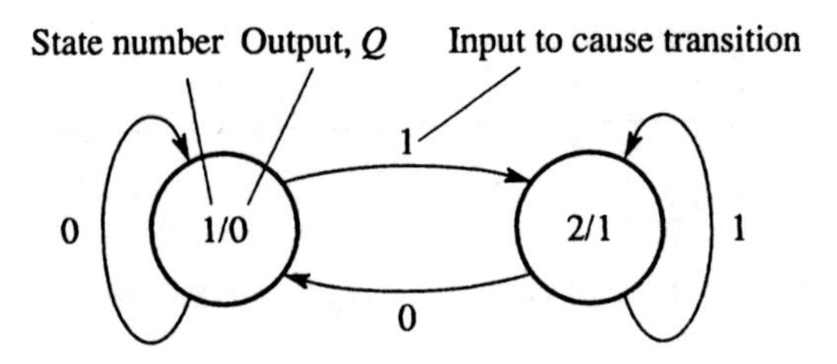

Figure 4.16 Moore model state diagram of the D-type flip-flop

cause a state change to state 2 ($Q = 1$). Similarly, if in state 2 ($Q = 1$), $D = 0$ will cause a state change to state 1. Some input values may not cause a state change, and in these cases, the arcs will return to the same state as "slings". In our flip-flop, $D = 0$ in state 1 does not cause a state change and $D = 1$ in state 2 also does not cause a state change. Figure 4.16 shows the corresponding slings.

An alternative type of state diagram is the *Mealy model state diagram* which will be introduced in Chapter 5. For complex circuits, it is usually necessary to use Boolean expressions which return the values on the arcs, rather than Boolean values themselves. This labelling is also introduced in Chapter 5.

4.3.3 J–K flip-flop

One of the most common requirements is to create a circuit whose outputs change from a 0 to a 1 or from a 1 to a 0, the so-called *toggle action.* A common example is the binary counter sequence described in Section 4.1. The *J–K* flip-flop provides this toggle operation in additional to being able to set the output to a 1 or reset the output to a 0. The truth table of the *J–K* flip-flop is shown in Figure 4.17.[1] The notation Q_- can be used to indicate the value of Q before the activating clock transition though we shall simply use Q. When $J = K = 0$, the output does not change. The J input cor-

J	K	Q_+
0	0	Q
0	1	0
1	0	1
1	1	$\overline{Q}$

Figure 4.17 Truth table of J–K flip-flop

[1] A flip-flop can be designed to only have a toggle action, the so-called *T* flip-flop. The single *T* input of this flip-flop corresponds to *J* and *K* connected together in a *J–K* flip-flop. *T* flip-flops are not usually found as packaged devices.

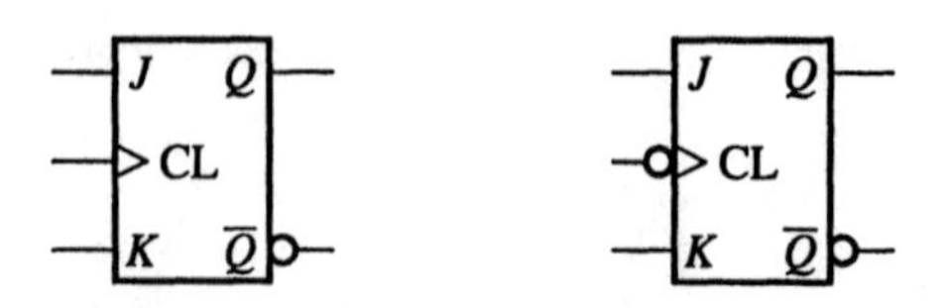

Figure 4.18 J–K flip-flop symbols

responds to the S input of an S–R flip-flop; when $J = 1$ (and $K = 0$), the output is set to a 1. The K input corresponds to the R input of an S–R flip-flop; when $K = 1$ (and $J = 0$), the output is set to a 0. The condition $J = K = 1$ is used to complement the output (toggle); if $Q = 0$, $Q_+ = 1$, and if $Q = 1$, $Q_+ = 0$.

The characteristic equation of the J–K flip-flop is:

$$Q_+ = J\overline{Q} + \overline{K}Q$$

which can be the basis of a J–K flip-flop design. A direct implementation of this equation will not work (see Wilkinson, 1992).

The symbols for positive edge-triggered and negative edge-triggered J–K flip-flops are shown in Figure 4.18. As with the D-type flip-flop, normally we shall use packaged J–K flip-flops in our design so that the internal implementation (which is actually an asynchronous circuit) is not significant. Of course, the external timing specification is important.

An example of a packaged TTL edge-triggered J–K flip-flop is the 74LS109 dual positive edge-triggered J–K flip-flop. (In the 74LS109, the K input has been inverted to $\overline{K}$.) Examples of dual negative edge-triggered J–K flip-flops include the 74LS73, 74LS76, 74LS78, 74LS112, 74LS113 and 74LS114. Larger numbers of J–K flip-flops in one package are not widely found. (Note that a J–K flip-flop has twice the number of "data" inputs of a D-type flip-flop which limits the number of J–K flip-flops that can be held in one package, given the package pin limitations.)

The Moore model state diagram of the J–K flip-flop is shown Figure 4.19. Notice

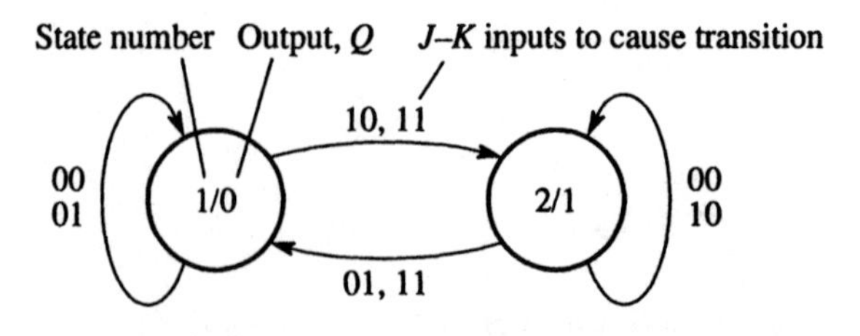

Figure 4.19 Moore model state diagram of the J–K flip-flop

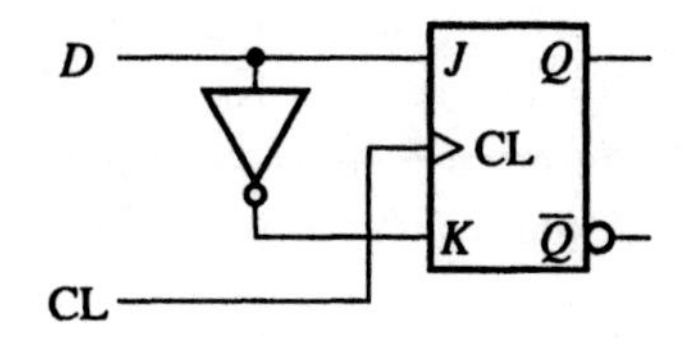

Figure 4.20 J–K flip-flop as a D-type flip-flop

that we now have two inputs and more than one condition to cause a state transition. For example, to change from state 1 ($Q = 0$) to state 2 ($Q = 1$) either $J = 1, K = 0$, or $J = 1, K = 1$.

A J–K flip-flop can emulate a D-type flip-flop simply by applying D to J and $\overline{D}$ to K using an inverter as shown in Figure 4.20 in a similar manner as is used to convert an S–R latch to a D latch.

SELF-ASSESSMENT QUESTION 4.4

How can a D-type flip-flop emulate a J–K flip-flop? (Clue: look at the J–K flip-flop characteristic equation.)

4.4 Registers

A flip-flop can store one binary digit. Often, we want to store a group of binary digits – a binary *word* – say 8 bits (a byte). To store eight bits, we can use eight flip-flops. The most convenient flip-flop for this application is the D-type flip-flop (though J–K flip-flops can be used as configured in Figure 4.20). Figure 4.21 shows eight D-type flip-flops formed into an 8-bit *register*. Such arrangements are packaged into one (small-scale) integrated circuit. Registers are very widely used inside computers for holding binary numbers.

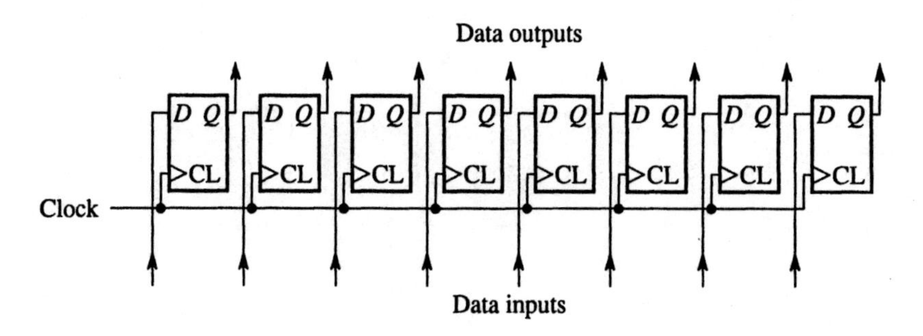

Figure 4.21 Eight-bit data register

Sometimes we need a way of setting the contents of registers to a specific value to initialize the system. This can be achieved using the asynchronous set and reset inputs (see Figure 4.15). To reset all outputs to zero all the reset inputs would be connected together and to a signal which causes the initialization.

SELF-ASSESSMENT QUESTION 4.5

How could you set an 8-bit register to the output pattern 010101010?

4.4.1 Shift registers

In a *shift register*, the outputs of each flip-flop pass onto the adjacent flip-flop by simply connecting the output of each flip-flop to the input of the adjacent flip-flop. In the design shown in Figure 4.22, a single data input is applied to the first flip-flop, and successive clock activations will cause the data to shift (move) one place right. Therefore, given an initial value of 00000000 in the shift register and a 1 applied to the first flip-flop, successive clock pulses will result in the values 10000000, 11000000, 11100000, 11110000, 11111000, 11111100, 11111110 and 11111111 in the register (assuming the 1 applied to the first flip-flop remains for each clock activation). If the 1 were removed (i.e. a 0 applied) immediately after the first clock activation, we would get: 10000000, 01000000, 00100000, 00010000, 00001000, 00000100, 00000010 and 00000001 in the register. And if *serial* data is applied (that is, a pattern of 0's and 1's applied to a single data input line one bit after another) this data could be entered into the register. The actual data entered would depend upon the logical value on the data line at the time of the activating clock transition.

Figure 4.22 is a *serial-in parallel-out shift register*; the information is entered "serially" (one bit at a time) at one end and is read out in "parallel" (all bits together) from the flip-flops. In the alternative *serial-in serial-out shift register*, data is entered serially at one end and retrieved serially from the last flip-flop, as shown in Figure 4.23. (A *parallel-in parallel-out register* would simply be a normal data register such as shown in Figure 4.21.)

A shift register can be formed using *J–K* flip-flops by connecting Q of one flip-flop to J of the next flip-flop, then $\overline{Q}$ of that flip-flop to K of the next flip flop, and so on as shown in Figure 4.24.

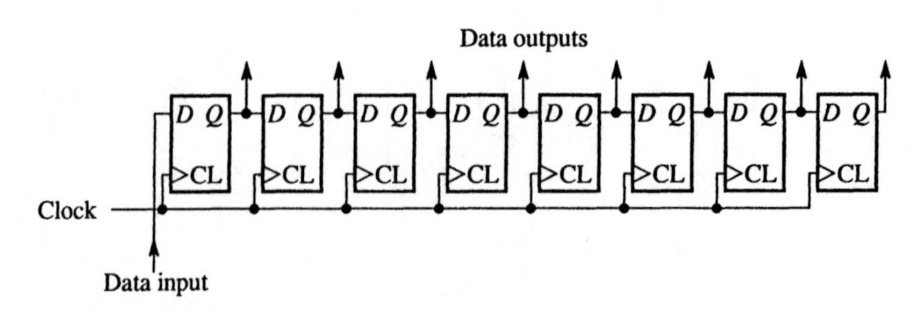

Figure 4.22 Eight-bit serial-in parallel-out shift register

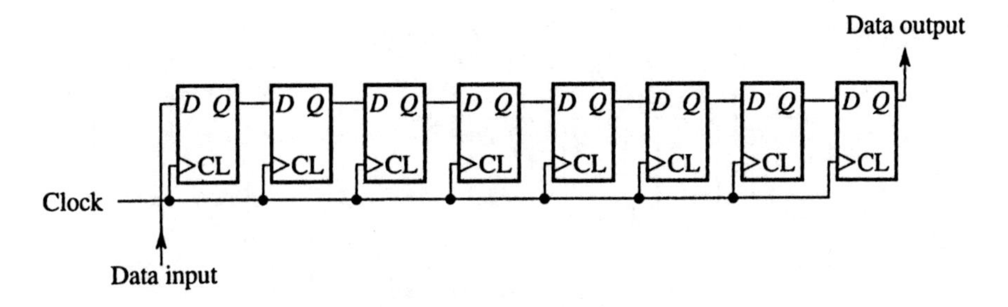

Figure 4.23 Eight-bit serial-in serial-out shift register

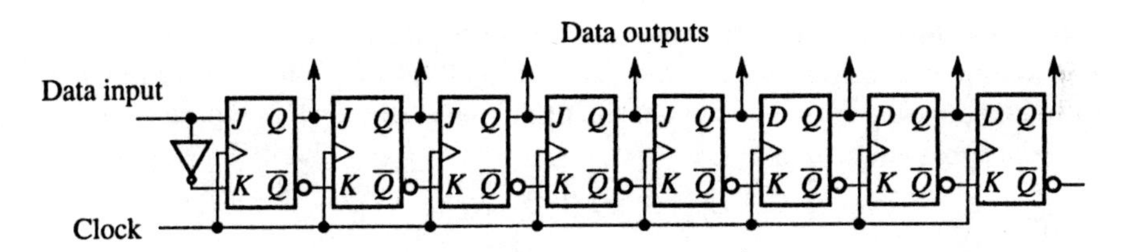

Figure 4.24 Eight-bit serial-in parallel-out shift register using J–K flip-flops

Applications

Shift registers find various applications in logic design. Here we shall mention some applications.

Multiplication and division

A register is often used to hold data, a set of bits that represents a binary number. A shift register can be used to divide the number by a power of 2 by simply shifting the number in the shift register the appropriate number of places right. For example, suppose the number 10 (decimal) is stored in an 8-bit register (i.e. 000001010 in binary). A one place right shift towards the most significant bit would result in the pattern 00000101, i.e. decimal 5.

> **SELF-ASSESSMENT QUESTION 4.6**
>
> What would another right shift of the pattern 00000101 produce?

If the shift register were organized to shift left (which requires the connections between the flip-flops to be from right to left rather than from left to right), the number could also be multiplied by a power of 2.

Generating control pulses

Shift registers find application in control circuits to generate sequences of logic

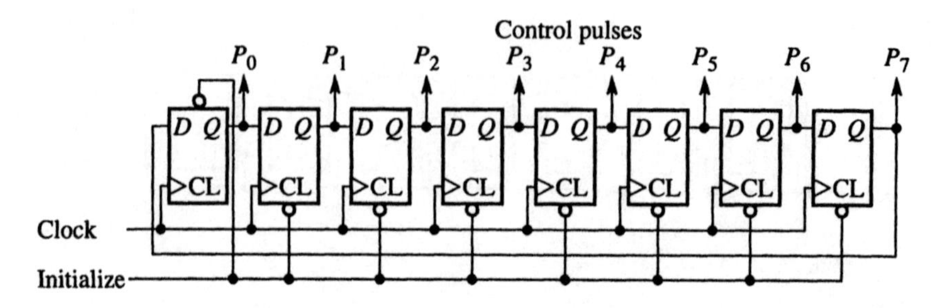

Figure 4.25 Ring counter with set/reset connections to initialize to 1000000

pulses. The final output of the shift register can be connected back to the first flip-flop to form a so-called *ring counter.* Suppose an 8-bit ring counter were initialized to 10000000 using asynchronous set and reset inputs, as shown in Figure 4.25. Then the repeating pattern:

10000000
01000000
00100000
00010000
00001000
00000100
00000010
00000001

would be generated by applying clock pulses. The Moore model state diagram is shown in Figure 4.26. Notice, for counters, that no inputs are shown on the state diagram as there are none except the clock input, and each transition will occur because of the clock transition only. Each output of the ring counter, P_0, P_1, P_2, P_3, P_4, P_5, P_6 and P_7, generates a 1 pulse for the duration of one clock cycle at different times as shown in Figure 4.27. These P pulses could be used to operate various parts of a digital system – for example, the internal workings of part of a computer.

If the final output in Figure 4.25 is complemented before returning to the first flip-

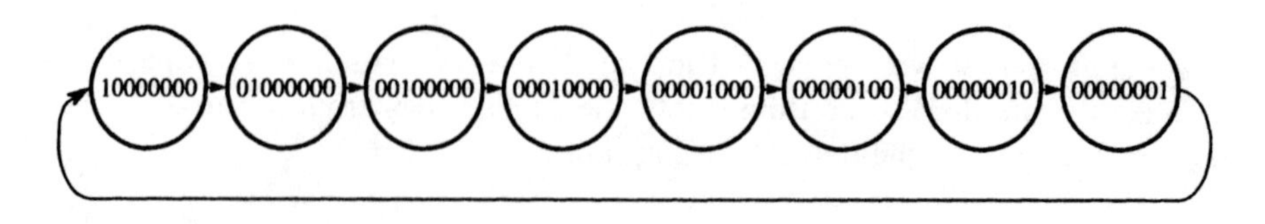

Figure 4.26 State diagram of ring counter

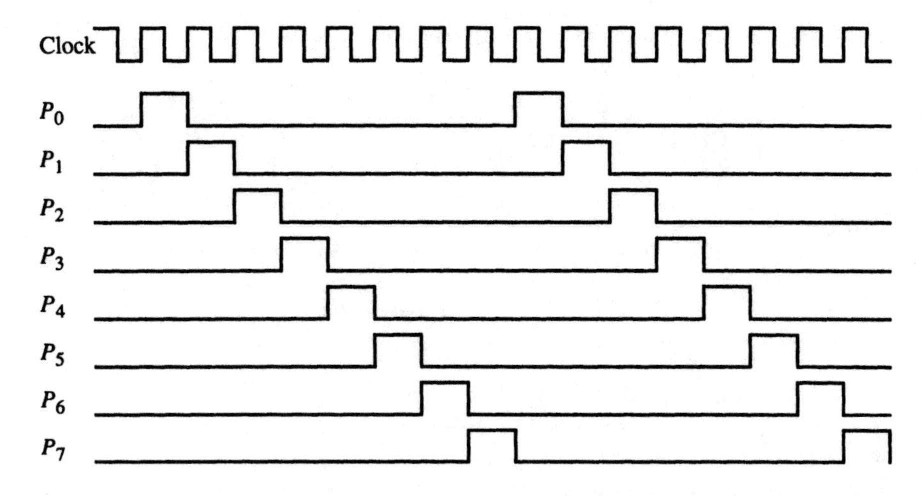

Figure 4.27 Generating control pulses with a ring counter

flop, we obtain a *twisted ring counter* (also called a *Johnson counter*) as shown in Figure 4.28.

The twisted ring counter, once initialized to zero, has a convenient repeating sequence: 00000000, 10000000, 11000000, 11100000, 11110000, 11111000, 11111100, 11111110, 11111111, 01111111, 00111111, 00011111, 00001111, 00000111, 00000011, 00000001, 00000000, … . In this sequence, only one bit changes from one number to the next. This is significant when using the outputs in decode circuits. For example, we could use one eight-input AND gate to recognize each pattern and get sixteen P pulses. When a number changes to the next, delays in the flip-flops will not cause other patterns to arise transiently, as might happen if more than one output had to change from one number to the next. The logic expressions for the P pulses are:

$$P_0 = \bar{D}_0\bar{D}_1\bar{D}_2\bar{D}_3\bar{D}_4\bar{D}_5\bar{D}_6\bar{D}_7 \qquad P_8 = D_0D_1D_2D_3D_4D_5D_6D_7$$
$$P_1 = D_0\bar{D}_1\bar{D}_2\bar{D}_3\bar{D}_4\bar{D}_5\bar{D}_6\bar{D}_7 \qquad P_9 = \bar{D}_0D_1D_2D_3D_4D_5D_6D_7$$
$$P_2 = D_0D_1\bar{D}_2\bar{D}_3\bar{D}_4\bar{D}_5\bar{D}_6\bar{D}_7 \qquad P_{10} = \bar{D}_0\bar{D}_1D_2D_3D_4D_5D_6D_7$$
$$P_3 = D_0D_1D_2\bar{D}_3\bar{D}_4\bar{D}_5\bar{D}_6\bar{D}_7 \qquad P_{11} = \bar{D}_0\bar{D}_1\bar{D}_2D_3D_4D_5D_6D_7$$
$$P_4 = D_0D_1D_2D_3\bar{D}_4\bar{D}_5\bar{D}_6\bar{D}_7 \qquad P_{12} = \bar{D}_0\bar{D}_1\bar{D}_2\bar{D}_3D_4D_5D_6D_7$$
$$P_5 = D_0D_1D_2D_3D_4\bar{D}_5\bar{D}_6\bar{D}_7 \qquad P_{13} = \bar{D}_0\bar{D}_1\bar{D}_2\bar{D}_3\bar{D}_4D_5D_6D_7$$
$$P_6 = D_0D_1D_2D_3D_4D_5\bar{D}_6\bar{D}_7 \qquad P_{14} = \bar{D}_0\bar{D}_1\bar{D}_2\bar{D}_3\bar{D}_4\bar{D}_5D_6D_7$$
$$P_7 = D_0D_1D_2D_3D_4D_5D_6\bar{D}_7 \qquad P_{15} = \bar{D}_0\bar{D}_1\bar{D}_2\bar{D}_3\bar{D}_4\bar{D}_5\bar{D}_6D_7$$

SELF-ASSESSMENT QUESTION 4.7

What would the sequence of an 8-bit twisted ring counter if it were initialized to 01010101?

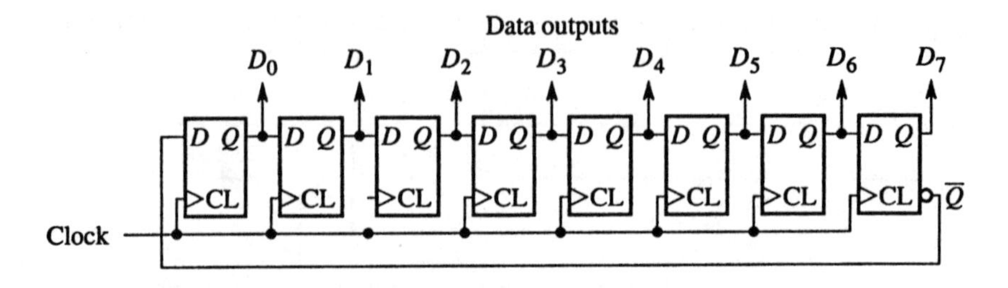

Figure 4.28 Twisted ring counter

SELF-ASSESSMENT QUESTION 4.8

How can twisted ring counter be implemented using *J–K* flip-flops?

Serial operations – historical

In Chapter 3, we encountered parallel addition in which pairs of digits were added together with separate full adders. An alternative solution to adding two binary numbers is to store them in shift registers and add pairs of numbers from these shift register in sequence using a single full adder, as shown in Figure 4.29. Notice the carry from one addition is used in the next addition. This solution would be *much* slower that using a parallel adder but is more economical of components. Only one full adder is required rather than *n* full adders with *n*-digit numbers (allowing for subtraction). Some early computers in the 1960s used serial adders.

4.5 Counters

A counter is a sequential circuit that will create a specific recurring output sequence. We have already mentioned the counting sequences in shift registers in the previous section. Counters are constructed with flip-flops with one flip-flop for each bit of the counting sequence. The most flexible flip-flop for counters is the *J–K* flip-flop

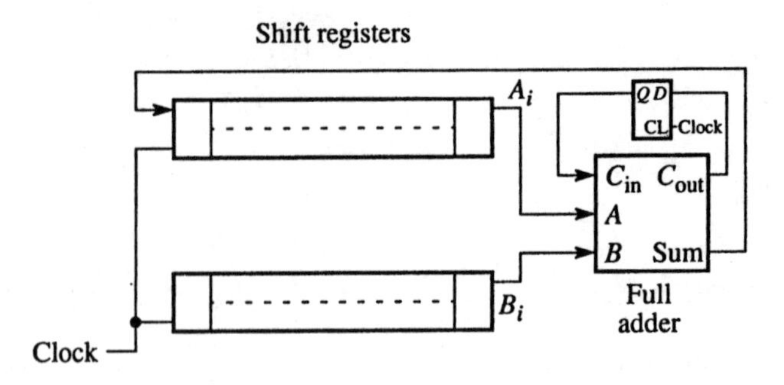

Figure 4.29 Serial adder (historical)

because that flip-flop can be made to complement its outputs with the application of appropriate J and K inputs (i.e. $J = K = 1$), in addition to being able to set or reset the outputs ($J = 1, K = 0$, and $J = 0, K = 1$ respectively). The D-type flip-flop can be used but often extra gates are needed to obtain the counting sequence. However, even though the D-type flip-flop is not the most natural choice, it may be that the final design is preferred, especially in VLSI, because the basic D-type flip-flop requires less internal transistors than a J–K flip-flop. Also the final design may require less interconnections than a J–K flip-flop solution. However, let us first consider the J–K flip-flop implementation, and then the D-type flip-flop implementation.

4.5.1 Counters using J–K flip-flops

To achieve synchronous operation – that is, getting all the required simultaneous output changes to occur together – all the flip-flops must be driven by the clock directly as shown in Figure 4.30. Then the problem reduces to determining what logical function must be applied to each flip-flop input to create the required sequence. Let us address this problem in a formal way which enables any binary sequence to be generated, using a simple 3-bit binary up sequence as our example. This sequence is shown as a table and as a state diagram in Figure 4.31, where A is the least significant bit and C is the most significant bit. The required flip-flops are also shown producing the outputs A, B and C. The output A is the least significant bit of the sequence and C the most significant bit of the sequence.

Finding the flip-flops' input functions

Given three flip-flops, we have three pairs of functions to determine: (J_A, K_A), (J_B, K_B) and (J_C, K_C). There are four possible scenarios for each flip-flip output (which applies to any counter):

1. No change in a 0 output
2. No change in a 1 output
3. A change from a 0 to a 1
4. A change from a 1 to a 0.

The required J/K inputs to effect each of the above are:

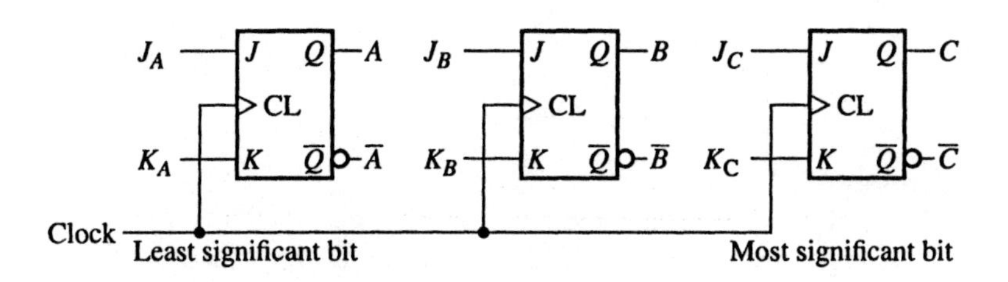

Figure 4.30 Inputs for a 3-bit counter using J–K flip-flops

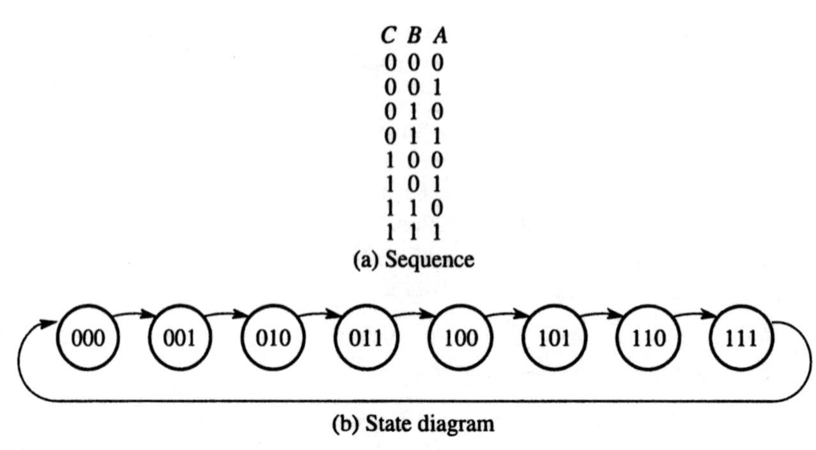

Figure 4.31 Binary (up-) counter sequence and state diagram

1. No change in a 0 output – $J = 0, K = 0$ or $J = 0, K = 1$
2. No change in a 1 output – $J = 0, K = 0$ or $J = 1, K = 0$
3. A change from a 0 to a 1 – $J = 1, K = 1$ or $J = 1, K = 0$
4. A change from a 1 to a 0 – $J = 1, K = 1$ or $J = 0, K = 1$

Notice there are two possible input values for each the above. For example, $J = 0$, $K = 0$ will result in no change irrespective of the value of the Q output, and appears in two entries, no change in a 0 and no change in a 1. Similarly $J = 0$, $K = 1$ will result in a 0 output irrespective of the initial output value, and appears in two entries, no change in a 0, and a change from a 1 to a 0.

Excitation table

We can reduce the conditions by using don't cares, as shown in Table 4.1. This table is known as the flip-flop *excitation table.*

Table 4.1 *J–K flip-flop excitation table*

$Q \rightarrow Q_+$	J	K
$0 \rightarrow 0$	0	X
$1 \rightarrow 1$	X	0
$0 \rightarrow 1$	1	X
$1 \rightarrow 0$	X	1

State (transition) table

Table 4.1 can be used to deduce the required Boolean functions of each flip-flop input. Let us first construct a *state table* (sometimes called a *state transition table*). A state table lists each state and the next state(s) and can be derived from the state diagram. The state table for our counter is shown in Table 4.2. If we look at entry A,

Table 4.2 *State table of binary counter in Figure 4.31*

Present state			*Next state*		
C	B	A	C_+	B_+	A_+
0	0	0	0	0	1
0	0	1	0	1	0
0	1	0	0	1	1
0	1	1	1	0	0
1	0	0	1	0	1
1	0	1	1	1	0
1	1	0	1	1	1
1	1	1	0	0	0

it so happens that there are no instances when A must remain at a 0 ("0 → 0") or remain at a 1 ("1 → 1"). It must change from a 0 to a 1 ("0 → 1") when the outputs CBA are 000, 010, 100 and 110 and at the next clock activation. According to the J–K excitation table, Table 4.1, this can be achieved by setting $J_A = 1$ on the input (K_A is a don't care and could be either 0 or 1). It must change from a 1 to a 0 ("1 → 0") when the outputs CBA are 001, 011, 101 and 111. According to the J–K excitation table, Table 4.1, this can be achieved by setting $K_A = 1$ on the input (J_A is a don't care and could be either 0 or 1). Hence we have the function for J_A and K_A in terms of C, B and A. A convenient approach is to map the function directly onto Karnaugh maps, which leads to the maps in Figure 4.32[2] for all flip-flop input functions:

$$J_A = K_A = 1$$
$$J_B = K_B = A$$
$$J_C = K_C = AB$$

The final counter circuit is given in Figure 4.33. Let us summarize the steps taken:

1. Draw the state (transition) table.
2. Draw the Karnaugh maps for each flip-flop input.
3. Derive the minimized flip-flop input functions.
4. Draw the final counter logic circuit.

This procedure can be used to design a counter to follow any sequence. For example, we can continue with the design for any number of variables. For four variables, A, B, C and D, we find that $J_D = K_D = ABC$. In fact, by examining the

---

[2] The order of A, B and C has been reversed on the Karnaugh maps because A is the least significant bit, but the minimized functions have been written in normal ABC order.

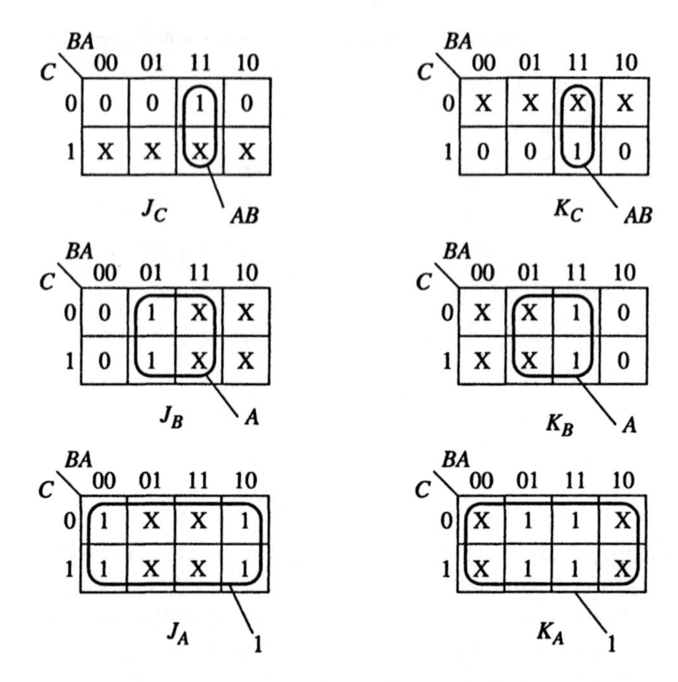

Figure 4.32 J–K input functions on Karnaugh maps for binary counter

binary sequence we can deduce the required input functions. Each output changes from a 0 to a 1 or from a 1 to a 0 (i.e. *toggles*) when all the lesser significant outputs are at a 1. The binary input function to both J and K inputs must be a 1 when a toggle action is necessary and is simply the logical AND of the lesser significant bits. This is demonstrated in Table 4.3 (overleaf). The input equations become:

$J_A = K_A = 1$ $\qquad$ $J_D = K_D = ABC$
$J_B = K_B = A$ $\qquad$ $J_E = K_E = ABCD$
$J_C = K_C = AB$ $\qquad$ $J_F = K_F = ABCDE$

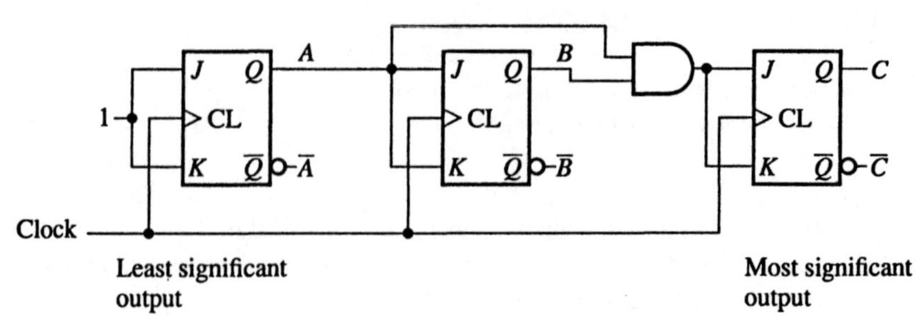

Figure 4.33 Three-bit synchronous binary counter design

Table 4.3 *Toggle actions in a synchronous binary-up sequence*

Clock pulse	*D*	*C*	*B*	*A*	
Φ_0	0	0	0	0	
Φ_1	0	0	0	1	toggle
Φ_2	0	0	1	0	
Φ_3	0	0	1	1	toggle
Φ_4	0	1	0	0	
Φ_5	0	1	0	1	toggle
Φ_6	0	1	1	0	
Φ_7	0	1	1	1	toggle
Φ_8	1	0	0	0	
Φ_9	1	0	0	1	toggle
Φ_{10}	1	0	1	0	
Φ_{11}	1	0	1	1	toggle
Φ_{12}	1	1	0	0	
Φ_{13}	1	1	0	1	toggle
Φ_{14}	1	1	1	0	
Φ_{15}	1	1	1	1	toggle
Φ_{16}	0	0	0	0	
		⋮			

Unfortunately, this design will lead to AND gates with large numbers of inputs for a large counter. An n-bit counter will require an AND gate with $n - 1$ inputs for the most significant stage. This requirement can be reduced by factorizing the Boolean expressions. Figure 4.34 shows one such design where only two-input AND gates are used. The problem with such designs is that there is now an increasing delay caused by the gates from the least significant output through to the most significant output. Each flip-flop cannot be activated until all the outputs prior to it have passed through a series of gates. This will severally limit the frequency at which the clock activations can be applied.

SELF-ASSESSMENT QUESTION 4.9

Modify the *J–K* flip-flop input equations for an 8-bit binary counter to use AND gates with a maximum of three inputs. Draw the gates.

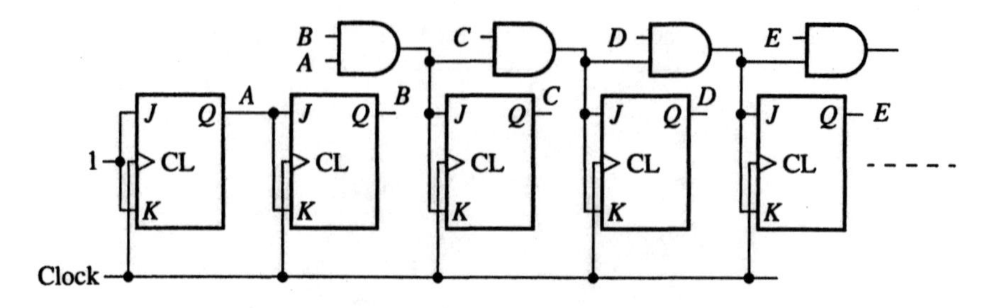

Figure 4.34 Binary counter using 2-input AND gates

4.5.2 Counters using D-type flip-flops

Let us now look at how *D*-type flip-flops can be used for a binary counter. The same procedure that was used for *J–K* flip-flops can be used for any type of flip-flop. Of course, the flip-flop excitation table must be used that matches the flip-flop. Table 4.4 shows the excitation table of the *D*-type flip-flop. There is only one input and no don't cares. In fact, *D* simply needs to be the new value of the *Q* output: a 0 if the output is to be a 0, or 1 if the output is to be a 1, irrespective of the original value of the output.

Table 4.4 *D-type flip-flop excitation table*

$Q \rightarrow Q_+$	D
$0 \rightarrow 0$	0
$1 \rightarrow 1$	1
$0 \rightarrow 1$	1
$1 \rightarrow 0$	0

Mapping the flip-flop input functions becomes a simple matter of listing the next output value for each output as shown in Figure 4.35. This yields the input functions:

$$D_A = \bar{A}$$
$$D_B = A\bar{B} + \bar{A}B$$
$$D_C = \bar{A}C + \bar{B}C + AB\bar{C}$$

and the logic circuit shown in Figure 4.36. Notice that the first (least significant) stage has a feedback connection from $\bar{Q}$ to *D*, which creates a toggle action as required. The other input functions are more complex than for the *J–K* flip-flop implementation in this particular case, and in general *D*-type flip-flops will require more complex input logic. However, as we have noted, the basic flip-flop itself is less complex, and there are less inputs which may reduce the wiring.

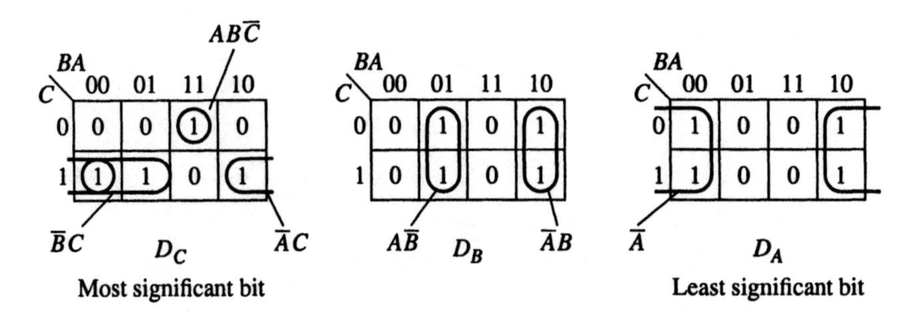

Figure 4.35 D input functions on Karnaugh maps for binary counter

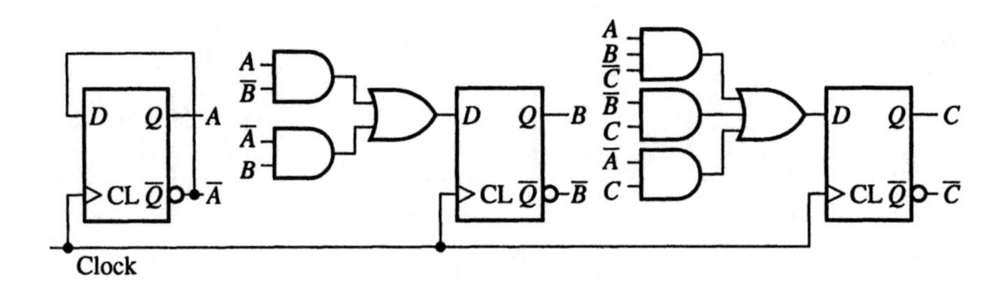

Figure 4.36 Synchronous binary counter using D-type flip-flops

The design techniques described can be applied for designing a counter to follow any recurring sequence. We shall give a further example in the next section and also in Chapter 5.

4.5.3 Start-up conditions

When power is applied to a logic system, the outputs of flip-flops are usually undefined. In a counter, what happens to the outputs when clock activations occur will depend upon the initial output values and the input functions. The sequence will depend upon the input functions. In some designs, not all output patterns occur in the required sequence, and it is possible for a completely new sequence to appear that does not include any of the patterns in the required sequence. This will depend upon how the don't cares were used in minimizing the input functions. If one of the required patterns appears, the counter will revert to the required sequence.

Start-up problems can be solved by either mapping outputs so that undefined output patterns will change to one of the defined output patterns in the sequence at some activating clock transition, or by using asynchronous reset/set inputs on the flip-flops to force the outputs to the required start-up values.

Let us demonstrate the first approach by designing a counter that follows the

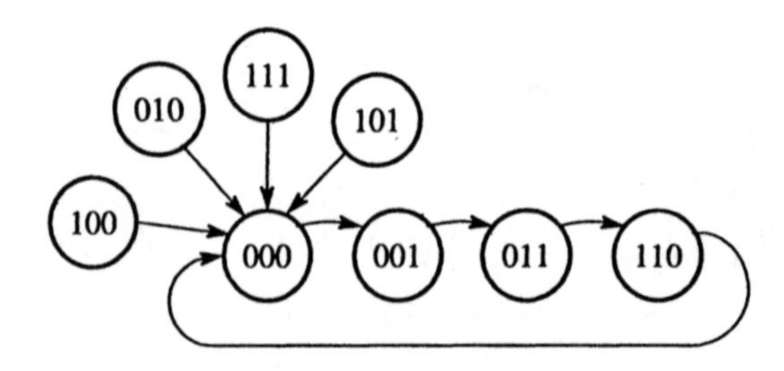

Figure 4.37 Counter sequence with undefined patterns mapped to pattern 000

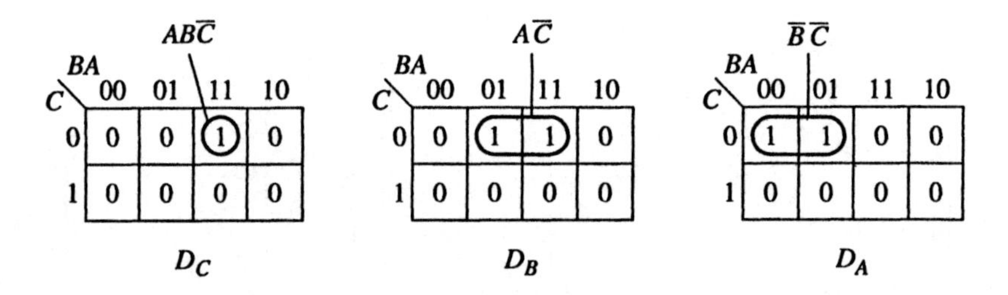

Figure 4.38 D-type flip-flop input functions

sequence 000, 001, 011, 110, 000, … . The state diagram is shown in Figure 4.37, together with mapping all undefined patterns to change to 000 on the next activating clock transition. The design using D-type flip-flops is shown in Figure 4.38 and the logic circuit in Figure 4.39.

4.5.4 Binary-down counters

A *binary-down counter* follows the sequence 111, 110, 101, 100, 011, 010, 001, 000, 111, … . To obtain a binary-down counter design, we can work through the counter design. Comparing the binary-up sequence with the binary-down sequence, we see

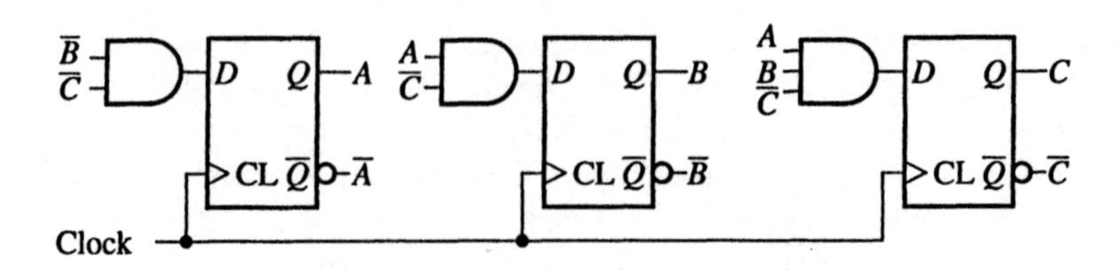

Figure 4.39 Counter implementation

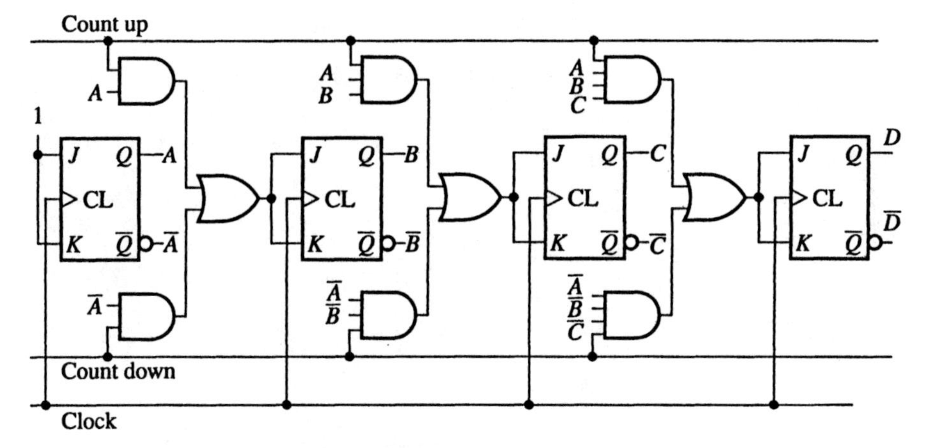

Figure 4.40 Bidirectional binary counter

that the down sequence can actually be obtained by simply inverting all the digits of the up sequence. Hence, by simply taking the $\overline{Q}$ outputs rather than the Q outputs we can obtain a binary-down counter. Alternatively, a binary-down counter can be created by using the complements of each output for the input functions, i.e. 1 (first flip-flop toggles in both counters), $\overline{A}$, $\overline{A}\,\overline{B}$, $\overline{A}\,\overline{B}\,\overline{C}$, ... This later approach would keep the outputs of the complete counter intact.

It follows from the previous design that a counter could be designed that would count up or count down, the so-called *bidirectional binary counter.* A binary control signal could be used to specify whether to count up or to count down, say a 0 to count up and a 1 to count down. The control signal would cause either Q or $\overline{Q}$ to be selected for the subsequent flip-flop input gates. The circuit arrangement is shown in Figure 4.40.

4.5.5 Ripple binary counters

There is a lower cost binary counter design if the output changes do not need to be in perfect synchronism; slight delays in output changes are acceptable. This design is the so-called *ripple counter* as signals have to ripple through from one end to the other. It is also called an *asynchronous binary counter*, though strictly the circuit is not an asynchronous sequential circuit. A clock still controls the overall counter but, internally, individual flip-flops are "clocked" from adjacent flip-flip outputs (apart from the first flip-flop). There may be slight delays before outputs change.

The binary-up ripple counter can be derived by examining the binary sequence as shown in Table 4.5. The least significant bit, A, toggles (changes from a 0 to a 1 or changes from a 1 to a 0) on every activating clock transition. The next least significant bit, B, toggles when A changes from a 1 to a 0. Similarly the next significant bit, C, toggles when B changes from a 1 to a 0. J–K flip-flops toggle by setting $J = K = 1$. D-type flip-flops can be made to toggle by connecting $\overline{Q}$ to the D input. Given

Table 4.5 *Toggle action for binary ripple counter*

Clock pulse	*C*	*B*	*A*
Φ_0	0	0	0
Φ_1	0	0	1
Φ_2	0	1	0
Φ_3	0	1	1
Φ_4	1	0	0
Φ_5	1	0	1
Φ_6	1	1	0
Φ_7	1	1	1
Φ_8	0	0	0
Φ_8	0	0	1
		⋮	

toggle

1 to 0

three flip-flops with outputs *A*, *B* and *C*, and flip-flops connected to toggle, we can cause *B* to toggle by connecting $\bar{A}$ to the clock input of *B* (assuming positive edge-triggered flip-flops). Similarly, we can cause *C* to toggle by connecting $\bar{B}$ to the clock input of *C*. The resulting binary-up counter based upon *D*-type flip-flops is shown in Figure 4.41. *J–K* flip-flops can be used in place of *D*-type flip-flops by simply applying 1's to both *J* and *K* inputs.

Examples of packaged TTL binary ripple counters include 74LS294 and 74LS197 4-bit asynchronous binary counters.

SELF-ASSESSMENT QUESTION 4.10

What connections are necessary to form a binary ripple counter if negative edge-triggered *D*-type or *J–K* flip-flops were used?

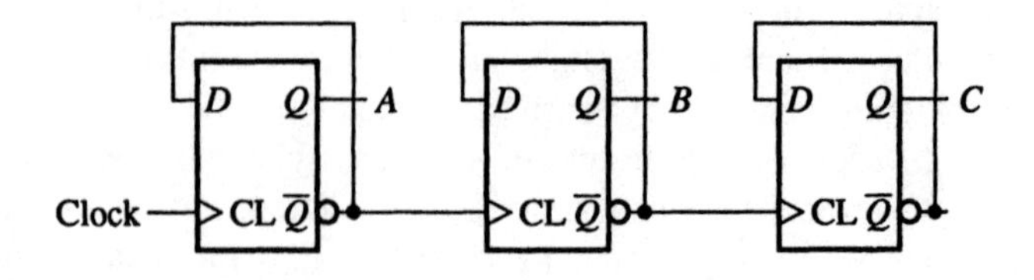

Figure 4.41 Binary ripple counter design

4.6 Summary

This chapter covered the following topics:

- The basic sequential circuit building blocks called flip-flops.
- The concept of a state diagram (Mealy model) and a state table.
- The formation of registers.
- Sequential circuits that follow binary sequences, called counters.
- The design of any synchronous counters having arbitrary sequences.

4.7 Tutorial questions

4.1 Determine the truth table of the flip-flop shown in Figure 4.42.

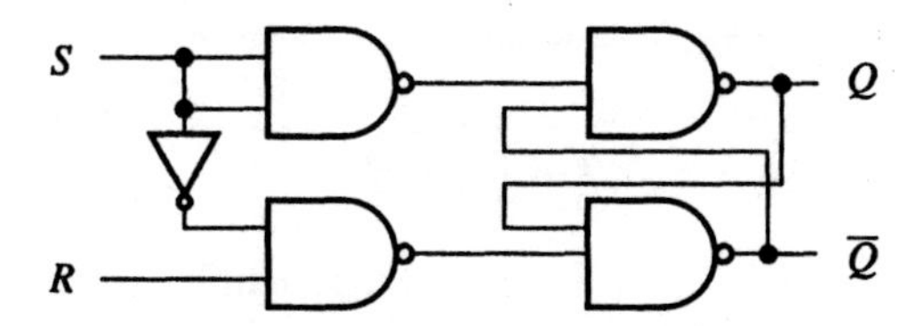

Figure 4.42 Circuit for Question 4.1

4.2 Complete the timing diagram shown in Figure 4.43 for a positive edge-triggered *J–K* flip-flop (i.e. give the *Q* output).

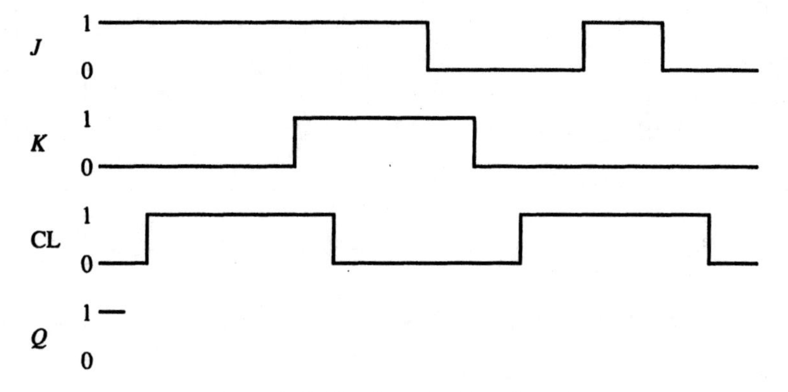

Figure 4.43 Waveforms for Question 4.2

4.3 What does a J–K flip-flop do when a series of activating clock transitions occur, if the initial Q output is a 0 and input functions are:

$$J = 1$$
$$K = \overline{Q}$$

4.4 Design a synchronous binary counter with outputs which follow a 3-bit Gray code as shown in the state diagram in Figure 4.44, using D-type flip-flops.

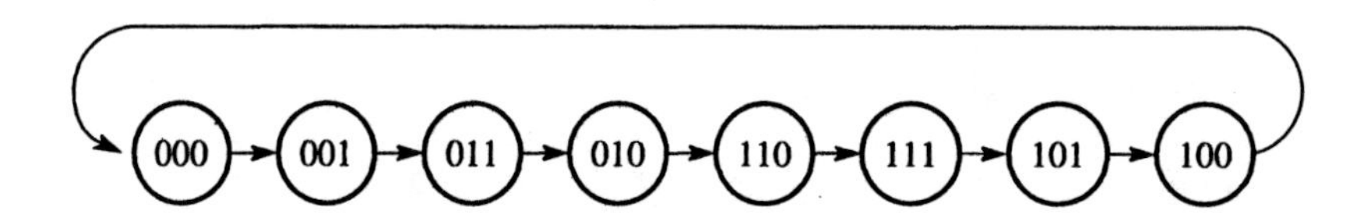

Figure 4.44 State diagram for Question 4.4

4.5 Repeat Question 4.4 using J–K flip-flops.

4.6 Design a binary counter which cycles repeatedly through the numbers 0 to 4 (i.e. binary 000 to 100). Ensure that the counter cannot cycle through any other sequence by mapping unused outputs to change to the pattern 000 after one activating clock transition. Use D-type flip-flops.

4.7 Repeat Question 4.6 using J–K flip-flops.

4.8 Determine the sequence followed by the counter shown in Figure 4.45.

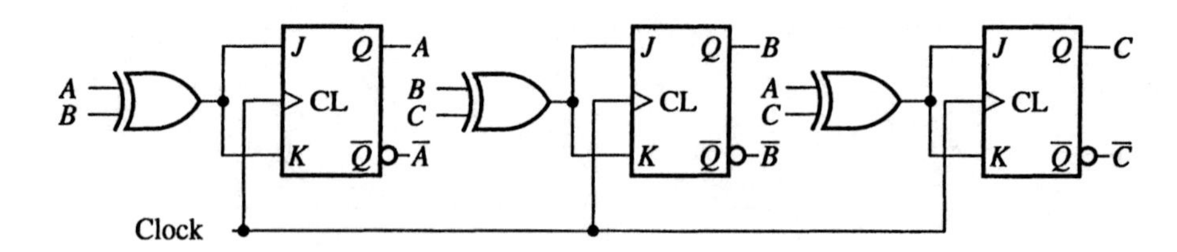

Figure 4.45 Counter for Question 4.8

4.9 Design a counter circuit which will implement the state diagram shown in Figure 4.46.

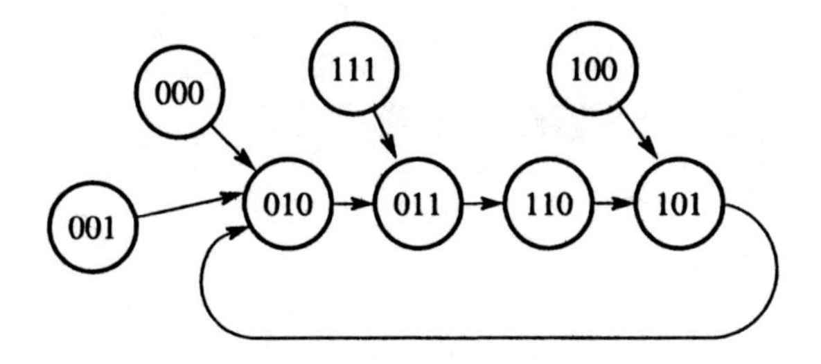

Figure 4.46 State diagram for Question 4.9

4.8 Suggested further reading

We have omitted the detailed internal design of J–K and D-type flip-flops to achieve the edge-triggered operation. Such details can be found in texts such as Wakerly (1994) and Katz (1994). These texts also include the master–slave flip-flop design.

Katz, R. H., *Contemporary Logic Design*, Benjamin/Cummings: Redwood City, California, 1994.

Wakerly, J. F., *Digital Design Principles and Practices, 2nd edition*, Prentice Hall: Englewood Cliffs, New Jersey, 1994.

CHAPTER 5

Sequential circuit design

Aims and objectives

In the previous chapter, we described the basic building blocks of sequential circuits, namely flip-flops, and showed how flip-flops can be used to create registers and counters. Now we shall describe a formal way of describing and designing any sequential logic circuit. Only considered are synchronous sequential circuits, in which the transition from one state to another occurs at the time of a clock transition. Various design problems are presented in this chapter to demonstrate the design methods.

5.1 Synchronous sequential circuit model

As we have seen in Chapter 4, a practical sequential circuit exists in a finite number of states. Hence such circuits are called *finite state machines* or simply *state machines*. Let us now develop a general model for a (synchronous) sequential circuit that can be a starting point for all such circuits. The states can be represented by binary signals called *state variables*. Each state has a unique combination of state variable values. The state variables are stored in flip-flops activated by the external clock signal. The model is shown in Figure 5.1 using D-type flip-flops.[1] This model is called the *Mealy model*. The circuit has a set of n inputs, labelled $X_{n-1} \ldots X_0$, and set of m outputs, $Z_{m-1} \ldots Z_0$. There are two sets of k internal signals, $y_{k-1} \ldots y_0$ and $Y_{k-1} \ldots Y_0$. The y signals define the present state of the circuit, i.e. the so-called *present state variables*, as stored in the flop-flops. The Y signals, called the *next state variables*, are generated to produce the next state of the circuit, and are inputs to the (state) flip-flops. The Y signals are combinational functions of the circuit inputs, $X_{n-1} \ldots X_0$, and the present state, $y_{k-1} \ldots y_0$. The circuit outputs, $Z_{m-1} \ldots Z_0$, are combinational functions of the inputs and the present state. To create a synchronous sequential circuit based upon this model, we shall need to work out the combinational functions for the next state variables, $Y_{k-1} \ldots Y_0$, and outputs, $Z_{m-1} \ldots Z_0$.

First, we have to establish how many flip-flops are needed for a particular design

[1] J–K flip-flops can be used but in that case there are twice the number of flip-flop inputs.

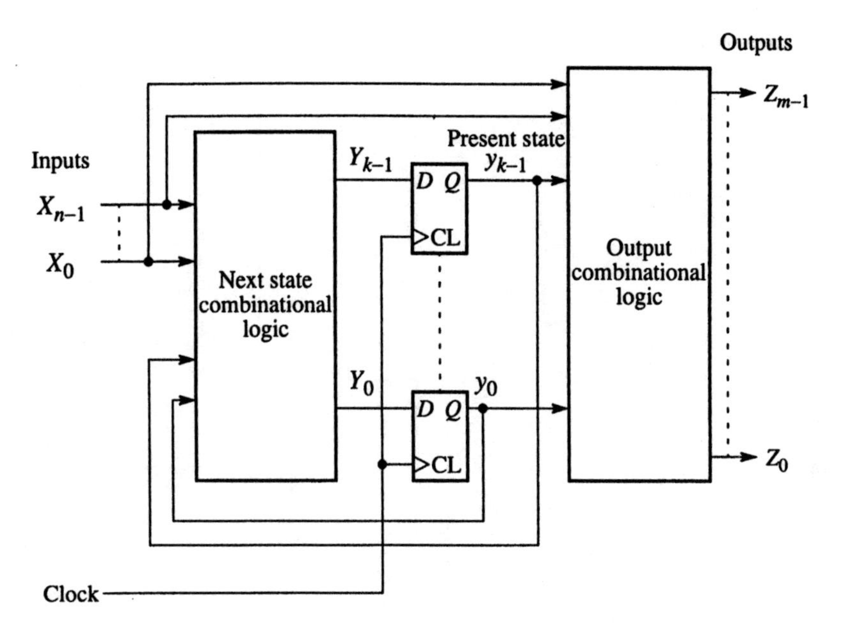

Figure 5.1 Synchronous sequential circuit model (Mealy model)

(i.e. k). If two different states are required, one state variable is needed which will be a 0 for one state and a 1 for the other state. If three or four states are required, two state variable are needed as four codes exist with two variables, 00, 01, 10 and 11, providing one code for each state. If between five and eight states are required, three state variables are needed, and so on. In general with between $2^{k-1} + 1$ and 2^k states, k state variables and hence k flip-flops are needed. If positive edge-triggered flip-flops are used, state changes occur on the rising edge of the clock signal. If negative edge-triggered flip-flops are used, state changes occur on the falling edge of the clock signal. In all cases, after one activating clock transition, sufficient delay must be allowed for signals to pass through the combinational logic circuits to establish next state values before the next activating clock transition.

Note that multiple changes in the (primary) inputs do not cause immediate changes in the state, as the change of state will be determined by the inputs and present state at the time of the activating clock transition (taking into account the delays in the circuit). Since the circuit outputs are combinational functions which include the inputs, changes in the inputs could cause the output to change immediately and not in synchronization with the clock. This problem can be avoided by making the outputs combinational functions of only the present state variables. This arrangement is embodied in the *Moore model* which we have already seen for flip-flops (Chapter 4, Section 4.3.2 and Section 4.3.3). For a general sequential circuit, the arrangement is shown in Figure 5.2.

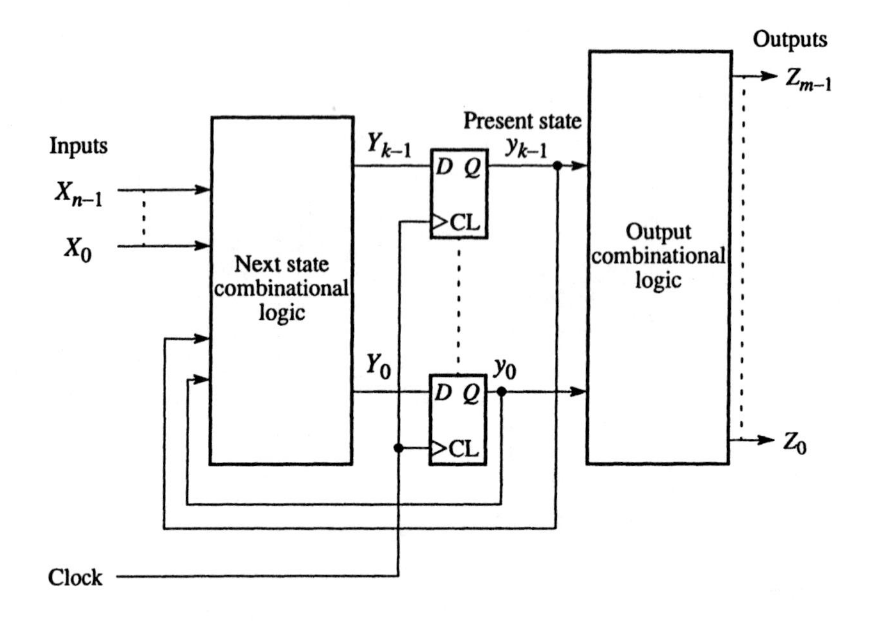

Figure 5.2 Synchronous sequential circuit model (Moore model)

A synchronous sequential logic circuit can often be designed starting from either a Mealy model or a Moore model, but if a Mealy model is used, the outputs may be affected by changes in the inputs immediately they occur rather than after a clock transition. Hence the behaviour of a Mealy model circuit and a corresponding Moore model circuit may be different in this respect. A counter is essentially a Moore model as the circuit outputs are directly from the flip-flops, i.e. $Z = y$.

SELF-ASSESSMENT QUESTION 5.1

How many flip-flops are needed for a state machine having forty-eight states?

5.2 Designing synchronous sequential circuits

Using the state machine model (either Mealy model or Moore model), there are two sets of functions to determine: the functions for the next state variables, $Y_{k-1} \ldots Y_0$, and the functions for the outputs, $Z_{m-1} \ldots Z_0$. Let us work through the procedure to obtain these functions by examples, starting with a Moore model design.

5.2.1 *Design problem 1 – arbitrary code counter with two sequences*

Suppose a synchronous counter is required which follows one of two repeating sequences, dependent upon a control input, C. When $C = 0$, the sequence is 00, 01, 11. When $C = 1$, the sequence is 00, 11, 01.

State diagram

The first step is usually to establish the state diagram derived from the problem specification. The current design problem naturally suits a Moore model as the outputs of the flip-flops (the state variables) are also the outputs of the circuit. The Moore model state diagram for our problem is shown in Figure 5.3. Only three states are in the sequence; the fourth state might appear upon switch-on, and hence we have mapped the fourth state to lead to state 1. In the state diagram, each state is identified by a state number followed by the corresponding outputs and each arc is marked with the value of C that will cause the transition. We have chosen the state numbers in a completely arbitrary manner. (Later, we consider a formal way of making this choice.)

State (transition) table

A state table is constructed from the state diagram, as shown in Table 5.1. (This state table is more accurately called a *state transition table* since it shows the transitions from one state to another state.) With four states, 1, 2, 3 and 4, we need two flip-flops, having outputs y_2 and y_1 (the present state variables).

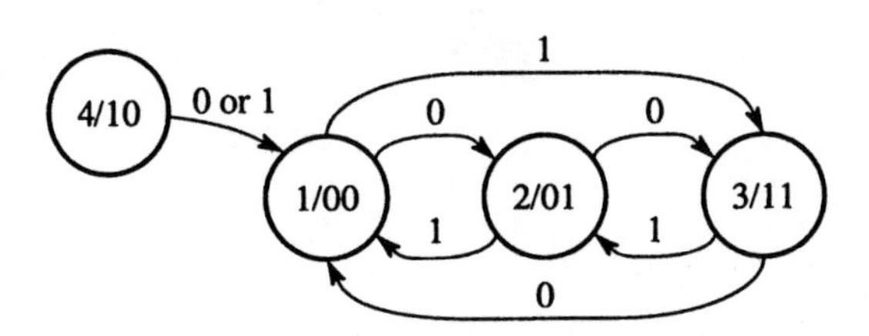

Figure 5.3 Moore model state diagram for counter problem

Table 5.1 *State table for counter problem*

Present state	*Next state*	
	C = 0	*C = 1*
1	2	3
2	3	1
3	1	2
4	1	1

State assignment

Values need to be chosen for the present state variables for each state. For convenience, we shall assign the state variable pattern $y_2y_1 = 00$ for state 1, $y_2y_1 = 01$ for state 2, $y_2y_1 = 11$ for state 3 and $y_2y_1 = 10$ for state 4, as these are also the required outputs. This "state assignment" leads to the so-called ***assigned state table*** shown in Table 5.2.

Table 5.2 *Assigned state table for counter problem*

Present state	*Next state* Y_2Y_1	
y_2y_1	$C = 0$	$C = 1$
00	01	11
01	11	00
11	00	01
10	00	00

Flip-flop input functions

Let us use *D*-type flip-flops, as in the model of Figure 5.2. The required inputs to the flip-flops, Y_1Y_2, can be established from the assignment state table. For *D*-type flip-flops, the *Y* inputs are simply the state values of the *y* outputs required after the next activating clock signal. (We are essentially using the *D*-type flip-flop excitation table as given in Chapter 4 – hence if *J–K* flip-flops were used, we would use the *J–K* flip-flop excitation table. Using *J–K* flip-flops is left as an exercise.)

The Karnaugh maps for the two inputs are shown in Figure 5.4, which gives the functions:

$$Y_2 = \bar{y}_2 y_1 \bar{C} + \bar{y}_2 \bar{y}_1 C$$
$$Y_1 = \bar{y}_2 \bar{C} + \bar{y}_2 \bar{y}_1 C + y_2 y_1 C$$

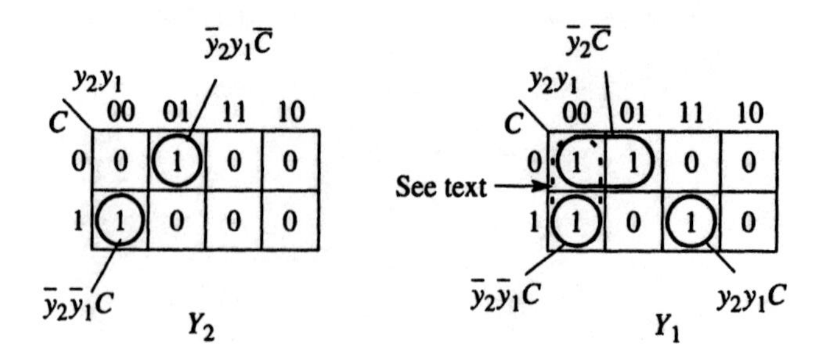

Figure 5.4 D-type flip-flop input functions

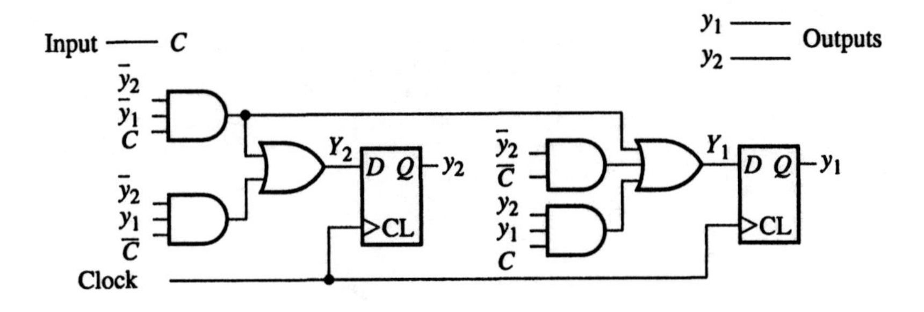

Figure 5.5 Counter circuit

Notice that Y_1 is not fully minimized, as would be achieved by choosing the dashed group instead of $\bar{y}_2\bar{y}_1C$. The term $\bar{y}_2\bar{y}_1C$ is chosen because this term also exists in Y_2, and hence one gate can be used for the term. The final logic circuit is shown in Figure 5.5.

SELF-ASSESSMENT QUESTION 5.2

Identify an alternative set of D-type flip-flop input equations from Figure 5.4 which share a term.

5.2.2 Design problem 2 – sequence detector

In the previous design, the state variables were also the circuit outputs. Let us now design a synchronous sequential circuit whose circuit outputs are some function of the state variables, and whose state encoding is not immediately obvious from the design specification. (It is clear with a counter that the state variables should follow the defined counter sequence.)

Consider a synchronous sequential logic circuit that will detect a defined serial pattern appearing on a single data input. The logic value on this input can change after each clock activation. The circuit has a single output which is a 1 when the defined sequence appears on the input. The general arrangement is shown in Figure 5.6 with one data input, x, and one output data output, Z, in addition to the clock input. Suppose the serial input to detect is 0110011, with Z becoming a 1 immediately after the last bit appears in the sequence. A circuit of this form might find application in several areas, including detecting the synchronization pattern recorded on magnetic disks before or after data, or synchronization patterns used in data communications. Sequence detectors also appear in other areas. For example, a combinational lock opens a door when a particular sequence of numbers is applied. If this were designed as a synchronous sequential circuit, a synchronizing clock signal would be necessary. (A combinational lock is essentially an asynchronous problem because the inputs can be altered at any time without a clock signal, and could be designed as such.)

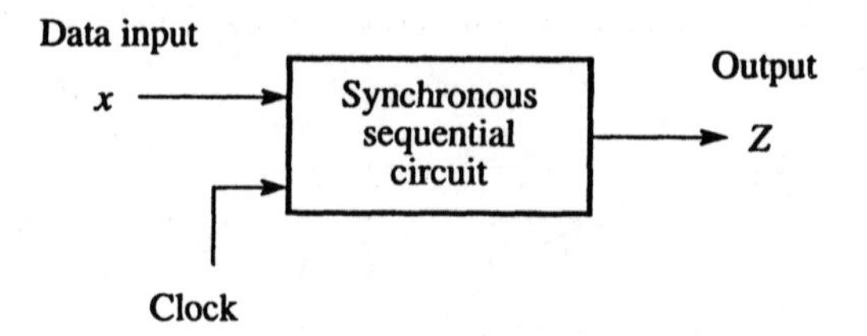

Figure 5.6 Sequence detector

Design procedure

The following are the design steps:

1. Derive the state diagram.
2. Draw the state table.
3. Assign state variable patterns to states.
4. Draw the assigned state table.
5. Derive the flip-flop input functions (on Karnaugh maps), and in our design:
6. Derive the output function of a Karnaugh map, and finally,
7. Draw the logic circuit.

State diagram

The state diagram can be derived from the problem specification. Our sequence has eight steps to reach the final pattern which suggests eight states, as shown in Figure 5.7. This is a Moore model state diagram in which each state has a specific output value. The initial state is state 1 with an output value of 0. The circuit changes from this state if a 0 appears on the input (or more accurately if a 0 is on the input at the time of the activating clock transition). This is the first 0 to be detected in the

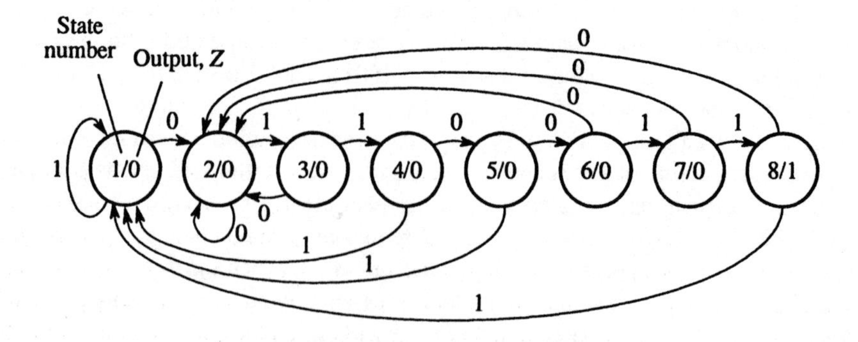

Figure 5.7 State diagram for sequence detector

sequence, and takes us to state 2. The output remains at a 0. Subsequent values of the required input pattern takes us to states 3, 4, 5, 6, 7, and finally to state 8. Only in state 8, when the final required bit of the input sequence is received, does the output change to a 1. After this, the next 1 takes us back to the initial state, state 1. If a 0 is received, it could be the first 0 of the sequence and hence we move to state 2 at this time. The other transitions in the state diagram before the final bit in the sequence is received either takes us to state 1 if a 1 is received or to state 2 if a 0 is received (as the first 0 of the sequence). In general, we have to consider all partial bit sequences before the final bit as possible sequences of the required pattern, unless there are other constraints in the design.

State (transition) table

The state table is shown in Table 5.3. In this state table, not only have the next states been listed but also the output values associated with each present state. In the Moore model that we are using here, each state has specific output values associated with it. In our current problem, there is a single output value, 0 or 1. We shall use the output column of the state table to determine the output function, but first let us make the state variable assignment.

Table 5.3 *State table of sequence detector*

Present state	*Next state*		*Output*
	$x = 0$	$x = 1$	Z
1	2	1	0
2	2	3	0
3	2	4	0
4	5	1	0
5	6	1	0
6	2	7	0
7	2	8	0
8	2	1	1

State variable assignment

Three state variables are needed, say $y_3y_2y_1$, for eight states,. The next step is to choose a state assignment, that is, suitable state variable values for each of the eight states. We could make this assignment in an arbitrary way. For example, state 1 = 001, state 2 = 010, state 3 = 011, state 4 = 100, state 5 = 101, state 6 = 110, state 7 = 111 and state 8 = 000. There is a very large number of possible assignments and each would lead to specific next state and output functions. We usually prefer an assignment that leads to the simplest functions, but unfortunately a general method that would work in all cases is not known. A simpler minimized function is obtained if the 1's on Karnaugh maps are adjacent, and we can try to achieve this by the following assignment, which we shall call "rule 1":

> **Rule 1** – Assign codes which differ in one variable to states that lead to the same next state.

Labelling three states that lead to the same next state using rule 1 is illustrated in Figure 5.8(a). Another rule, "rule 2", if the first cannot be done, is:

> **Rule 2** – Assign codes that differ in one variable for next states of a present state.

Rule 2 is illustrated in Figure 5.8(b). A final rule, "rule 3", if neither rule 1 nor rule 2 can be satisfied, is:

> **Rule 3** – Assign codes that differ in one variable to states with the same output (with the same inputs).

(a) Rule 1 – Assign codes which differ in one variable to states that lead to the same next state.

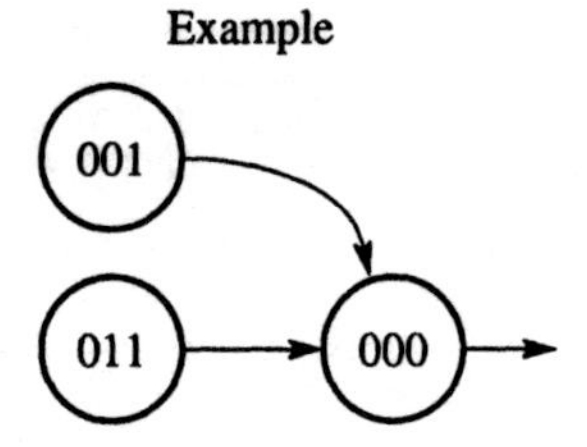

(b) Rule 2 – Assign codes that differ in one variable for next states of a present state.

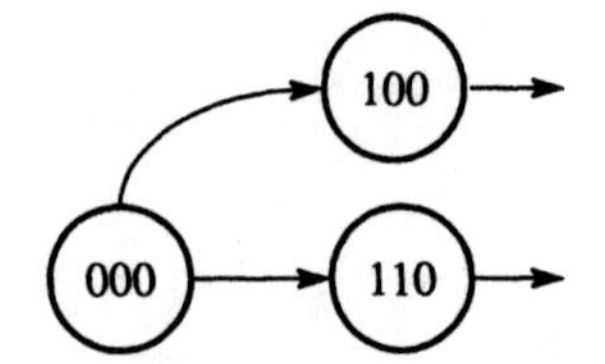

(c) Rule 3 – Assign codes that differ in one variable to states with the same output.

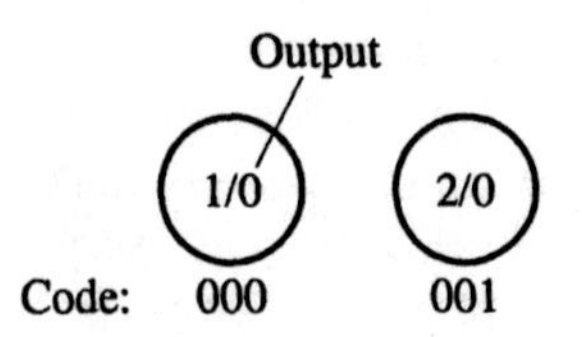

Figure 5.8 Labelling three states that lead to the same next state

Rule 3, illustrated in Figure 5.8(c), is intended to create the simplest output function. Rules 1 and 2 are intended to create the simplest next state functions. These rules are only suggestions and cannot guarantee a design with minimum logic.

The code called *Gray code* has the characteristic that adjacent codes differ by one bit. There are many possible Gray codes. A Gray code is used in Karnaugh map labelling. For three variables, the pattern of this Gray code is 000, 001, 011, 010, 110, 111, 101, 100. Hence a Gray code would be a suitable encoding for a series of states, where one state leads to one other state. For our problem, using rule 1 on states 4, 5 and 8 (states 4, 5 and 8 can lead to state 1), and using rule 1 on states 3, 6 and 7 (states 3, 6 and 7 can lead to state 2), we make the assignment:

State 1 = 000   State 2 = 111
State 4 = 001   State 3 = 101
State 5 = 010   State 6 = 110
State 8 = 100   State 7 = 011

Using 000 for state 1 enables the flip-flops to be initialized to state 1 easily. (Though not shown in the subsequent circuit, we assume the flip-flops have reset inputs that can be used to reset the outputs to 000.) State 8 can lead to either state 1 or state 2 but fits nicely with state 1. (It is left as an exercise to determine whether there are any other (non-equivalent) assignments, or whether using other rules would be better in this particular case.) The assigned state table is shown in Table 5.4.

Table 5.4 *Assigned state table of sequence detector*

Present state	*Next state,* $Y_3Y_2Y_1$		*Output*
$y_3y_2y_1$	$x = 0$	$x = 1$	Z
000	111	000	0
111	111	101	0
101	111	001	0
001	010	000	0
010	110	000	0
110	111	011	0
011	111	100	0
100	111	000	1

Flip-flop input functions

The input functions for the three flip-flops are obtained as before by mapping the next state variables in the assigned state table onto Karnaugh maps as shown in Figure 5.9 (assuming D-type flip-flops). The mapping is straightforward; we simply put the next

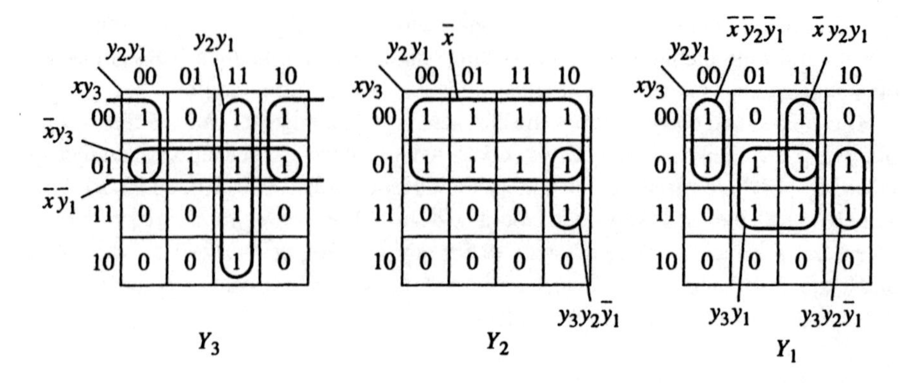

Figure 5.9 Input functions for sequence detector

state values on the maps which leads to the functions:

$$Y_3 = \bar{x}\bar{y}_1 + \bar{x}y_3 + y_2y_1$$
$$Y_2 = \bar{x} + y_3y_2\bar{y}_1$$
$$Y_1 = y_3y_1 + y_3y_2\bar{y}_1 + \bar{x}\bar{y}_2\bar{y}_1 + \bar{x}y_2y_1$$

Notice that the minimal solution for Y_1 is not chosen to allow gate sharing ($y_3y_2\bar{y}_1$).

SELF-ASSESSMENT QUESTION 5.3
What is the minimal function for Y_1?

Output function

The output function is obtained directly from the state table output column, and is mapped in Figure 5.10, giving the output function:

$$Z = y_3\bar{y}_2\bar{y}_1$$

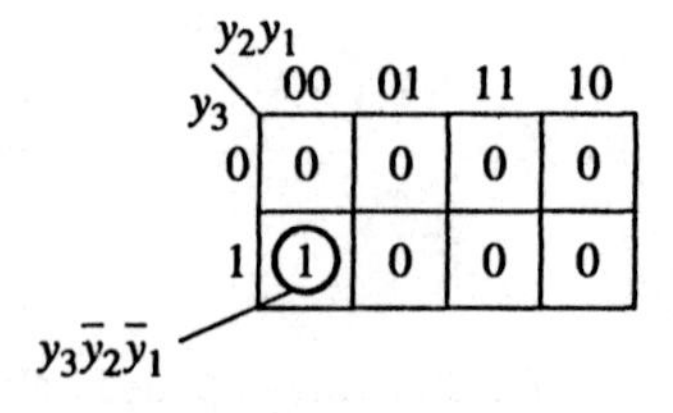

Figure 5.10 Output function, Z

Logic circuit

Finally, the logic circuit can be drawn from the functions, as shown in Figure 5.11.

SELF-ASSESSMENT QUESTION 5.4

Given four states, 1, 2, 3 and 4 in which 1 leads to 2, 2 leads to 3, 3 leads to 4, and 4 leads to 1, give two good different assignments of state variables assuming the first state is assigned 00.

5.2.3 *Using J–K flip-flops*

J–K flip-flops can be used instead of *D*-type flip-flops. We simply use the *J–K* flip-flop excitation table as given in Chapter 4, Section 4.5.1 to work out the next state functions from the state table. As mentioned in Chapter 4, this will generally lead to simpler functions than for a *D*-type flip-flop implementation, but as there are twice the number of inputs and hence functions, the logic could be more complex.

5.2.4 *Mealy model designs*

Recall that the Mealy model state machine differs from a Moore model in that each state does not necessarily have one set of output values. The outputs depend not only on the actual state but the inputs occurring to take it to that state. The Mealy model state diagram indicates the outputs on the arcs leading to the state, together with the inputs that caused the transition to that state. Often the Mealy model state diagram has less states than a Moore model state diagram for the same problem, and potentially less state variables and state flip-flops. However, the output function may be more complex as it will use the input variables as well as the state variables. Perhaps more importantly, the outputs could change if the inputs change between clock transitions. This may, or may not, be significant in a system design depending upon the application.

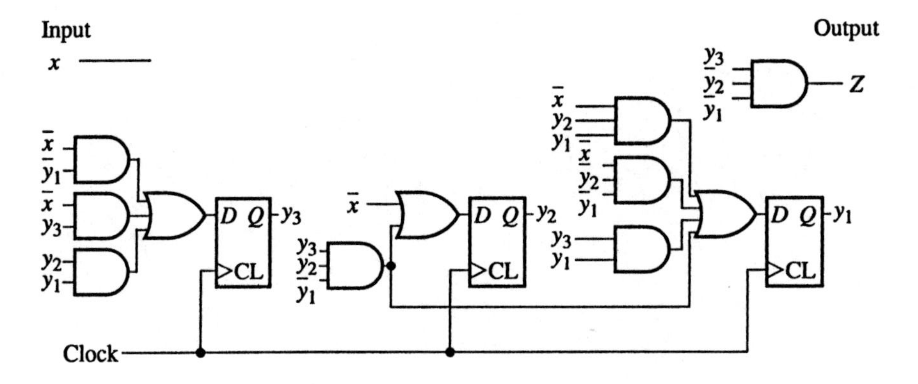

Figure 5.11 Sequence detector logic circuit

Let us look at a Mealy model solution to our sequence detector problem detecting the sequence 0110011. The same design steps have to be done, namely:

1. Derive the state diagram.
2. Draw the state table.
3. Assign state variable patterns to states.
4. Draw the assigned state table.
5. Derive the flip-flop input functions (on Karnaugh maps).
6. Derive the output function of a Karnaugh map, and finally:
7. Draw the logic circuit.

Mealy model state diagram

The Mealy model state diagram is shown in Figure 5.12. The difference between this state diagram and the Moore model state diagram is that state 8 is eliminated and the transition from state 7 to state 8 is replaced by a transition from state 7 to state 1 when the final bit of the sequence is received. Now the circuit could have one of two output values in state 1: either $Z = 0$ if the sequence has not been received, or $Z = 1$ if the sequence has been received.

Mealy model state table

The state table is shown in Table 5.5. Notice how the listed outputs are associated with the next states and for each input value, rather than the present states.

Mealy model assigned state table

Even though the number of states in the state diagram and table have been reduced from 8 to 7, we still need three variables to represent the states. Using the same assignment as for the Moore model, we get the assigned state table shown in Table 5.6.

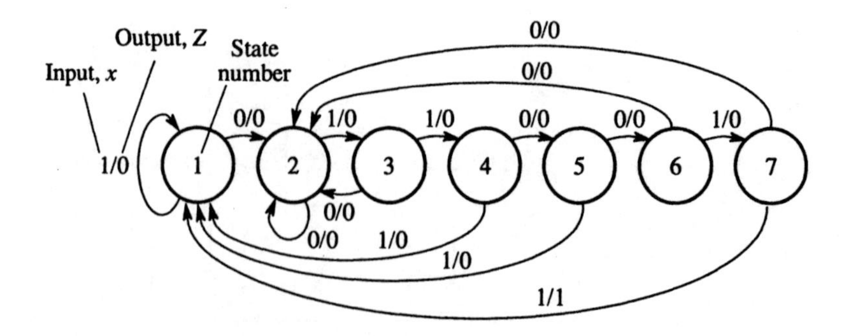

Figure 5.12 Mealy model state diagram for sequence detector

Table 5.5 *Mealy model state table of sequence detector*

Present state	*Next state*		*Next output, Z*	
	x = 0	*x = 1*	*x = 0*	*x = 1*
1	2	1	0	0
2	2	3	0	0
3	2	4	0	0
4	5	1	0	0
5	6	1	0	0
6	2	7	0	0
7	2	1	0	1

Table 5.6 *Assigned state table of sequence detector*

Present state	*Next state, $Y_3Y_2Y_1$*		*Next output, Z*	
$y_3y_2y_1$	*x = 0*	*x = 1*	*x = 0*	*x = 1*
000	111	000	0	0
111	111	101	0	0
101	111	001	0	0
001	010	000	0	0
010	110	000	0	0
110	111	011	0	0
011	111	000	0	1

Flip-flop input functions

As before, from the assigned table we can obtain the next state functions. One combination of variables, i.e. $y_3y_2y_1 = 100$, is not used and is entered on the Karnaugh maps as a "don't care". (We are assuming a general reset signal will force the flip-flops into 000 upon switch on, so that $y_3y_2y_1 = 100$ will never occur.) The Karnaugh maps of the input functions are shown in Figure 5.13 (to include shared groups). This leads to the functions very similar to previously:

$$Y_3 = \bar{x}\bar{y}_1 + \bar{x}y_3 + y_3y_2y_1 + \bar{x}y_2y_1$$
$$Y_2 = \bar{x} + y_3\bar{y}_1$$
$$Y_1 = y_3 + \bar{x}\bar{y}_2\bar{y}_1 + \bar{x}y_2y_1$$

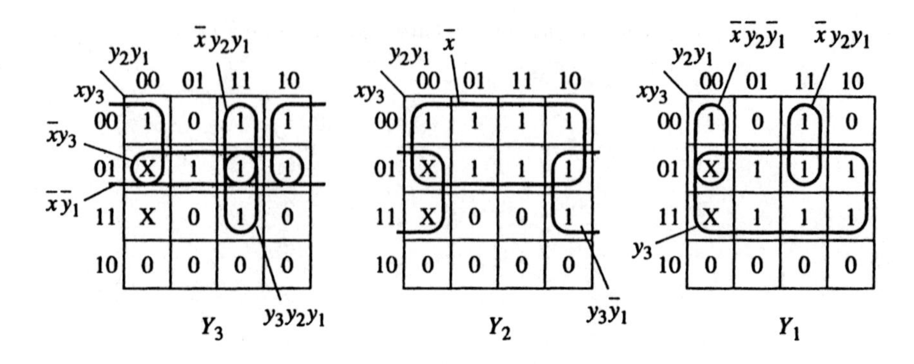

Figure 5.13 Input functions for Mealy model sequence detector

Output function

The output function will be a function of x and the present state variables, $y_3y_2y_1$, which can be taken from the assigned state table and mapped as shown in Figure 5.14, i.e.:

$$Z = x\bar{y}_3y_2y_1$$

The circuit follows from the equations.

SELF-ASSESSMENT QUESTION 5.5

What would happen, if anything, to the Z output if x changes from a 1 to a 0 while Z is a 0?

5.2.5 *State minimization*

The number of states in the state diagram will determine the number of state variables, and hence the number of flip-flops. If the number of states is between $2^{n-1} + 1$

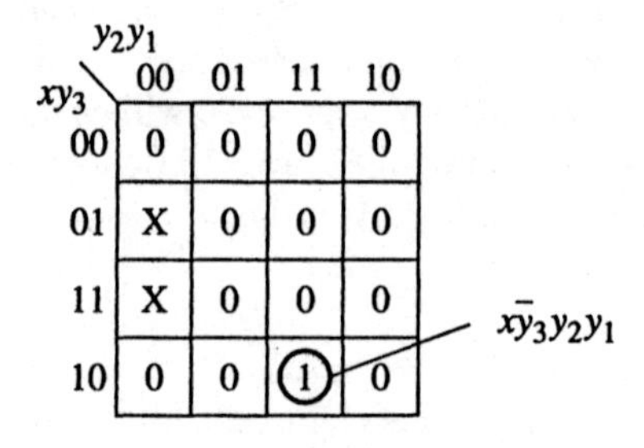

Figure 5.14 Mealy model output function, Z

and 2^n, the number of state variables and flip-flops is n (assuming fully encoding the states). Clearly, we would like to reduce the number of states to a minimum because that would reduce the circuitry. For simple problems, the state diagram, as derived from the problem specification, will probably already use the minimum number of states. However, for more complex problems, the intuitively created state diagram may be more complex than necessary, depending upon the skill of the designer. In any event, the original state diagram must be correct and it may that the correctness can be more easily seen from a larger than necessary state diagram.

In this section, let us look at how the number of states can be reduced. *Identical states*, i.e. states with identical next state and output entries in the state table, can certainly be merged into a single state with a single number. (Though unusual, identical states occur in an example in Section 5.2.7.) Next we need to examine states that can be made the same if other states are the same or were made the same. Such states are called *equivalent states*.

Equivalent states

The goal here is recognizing that two states are equivalent. Two states are *equivalent* and hence can be replaced by one state if the following two conditions exist:

1. The outputs[2] associated with the two states are the same, and
2. Corresponding next states are the same, or equivalent.

Once these two conditions are satisfied, we can combine the two states into one newly named state. The first condition immediately rules out pairs of states if the outputs are different. These states are called *output incompatible*. States must be output compatible for combining. Once pairs of states are eliminated that are output incompatible, we then look at all combinations of pairs that might satisfy the second condition.

Consider, for example, the Moore model state table shown in Table 5.7. In this table, there are two inputs to the circuit, x_1 and x_2, and a single output Z. No states are actually identical. States 3 and 7 have the same output value and are hence output compatible, but are output incompatible with all the other states, 1, 2, 4, 5 and 6. States 1, 2, 4, 5 and 6 are output compatible between themselves. So we now look at states 3 and 7 as possibly equivalent and pairs of states among 1, 2, 4, 5 and 6 as possibly equivalent.

"Possibly equivalent" pairs can be listed together with the next states that have to be equivalent in a table, as shown in Table 5.8. The required pairs have been taken directly from the state table. For example, to make states 1 and 2 equivalent, when $x_1x_2 = 10$, state 1 must be equivalent to state 4. (The other entries are the same or would be the same if $1 = 2$.) Continuing through the table we get a list of pairs that must be equivalent to make the "possibly" equivalent pair actually equivalent. Sometimes, the required pairs are output incompatible. We mark the final column if this is

[2] Present outputs for Moore model, next outputs for Mealy model.

Table 5.7 *Moore model state table with equivalent states*

Present state	*Next state*				*Output, Z*
	x_1x_2				
	00	*01*	*11*	*10*	
1	5	3	2	1	0
2	5	3	1	4	0
3	3	4	4	5	1
4	5	3	2	2	0
5	6	7	1	1	0
6	3	3	1	7	0
7	7	1	1	5	1

Table 5.8 *Equivalent pairs of states from Table 5.7*

Possibly equivalent pairs	*Required equivalent pairs*	*Output incompatible*
1, 2	(1, 2), (1, 4)	
1, 4	(1, 2)	
1, 5	(5, 6), (3, 7), (1, 2)	x
1, 6	(3, 5), (1, 2), (1, 7)	x
2, 4	(1, 2)	
2, 5	(5, 6), (3, 7), (1, 4)	x
2, 6	(3, 5), (4, 7)	x
3, 7	(3, 7), (1, 4)	
4, 5	(5, 6), (3, 7), (1, 2)	x
4, 6	(3, 5), (1,2), (2, 7)	x
5, 6	(3, 6), (3, 7), (1, 7)	x

so, because then the "possibly" equivalent pair cannot be equivalent. Once this is done, we go through the table again to see if any newly identified output incompatible pair are in the required list. If so, we mark them as incompatible. This process is repeated until we are left with only output compatible pairs.[3] In our case, we have the pairs (1, 2), (2, 4), (1, 4) and (3, 7) as output compatible and equivalent. States 5

[3] The procedure for finding equivalent pairs can be assisted by using a so-called *implication chart* (see Katz, 1994).

and 6 will not combine with any other state and remain. Since 1 ≡ 2, 2 ≡ 4 and 1 ≡ 4, we can say that 1 ≡ 2 ≡ 4. (Even had we only found say 1 ≡ 2 and 2 ≡ 4, we could still have deduced that 1 ≡ 2 ≡ 4; this may not be possible if the table has don't cares, see later.)

Four states are left, which could be renamed *A* (for states 1, 2 and 4), *B* (for states 3 and 7), *C* (for state 5) and *D* (for state 6). Making this substitution, we get the reduced state table shown in Table 5.9. The circuit can then be designed using this reduced state table, with only two state variables instead of three as would have been necessary initially.

Table 5.9 *Reduced Moore model state table*

Present state	*Next state*				*Output, Z*
	x_1x_2				
	00	*01*	*11*	*10*	
A	*C*	*B*	*A*	*A*	0
B	*B*	*A*	*A*	*C*	1
C	*D*	*B*	*A*	*A*	0
D	*B*	*B*	*A*	*B*	0

SELF-ASSESSMENT QUESTION 5.6

Find a suitable state assignment for Table 5.9.

Minimization of Mealy model state tables

The same minimization procedure used for Moore model state diagrams can be applied to a Mealy model state table. In this case, next state outputs must match in each column for the states to be output compatible.

5.2.6 State diagrams using transition expressions

If the transition from one state to another state is specified by several input variables, as would be the case in a complex digital system, the previously constructed state diagrams would be clumsy. Take the example of a circuit with three separate inputs, *a*, *b* and *c*. Suppose the transition from state 1 to state 2 occurs when abc = 001, but otherwise no state changes occur. In our notation we would have the diagram as shown in Figure 5.15(a). A more convenient notation would be to simply give the Boolean condition necessary for the transition or no transition in so-called *transition expressions*, as shown in Figure 5.15(b). Each transition expression can be reduced or written in any correct Boolean form. For example, $\overline{\bar{a}\bar{b}c}$ could be written as $a + b + \bar{c}$. If transition expressions are used, it is very important that all combinations of variable values are accounted for in the diagram. Also, a particular combination of

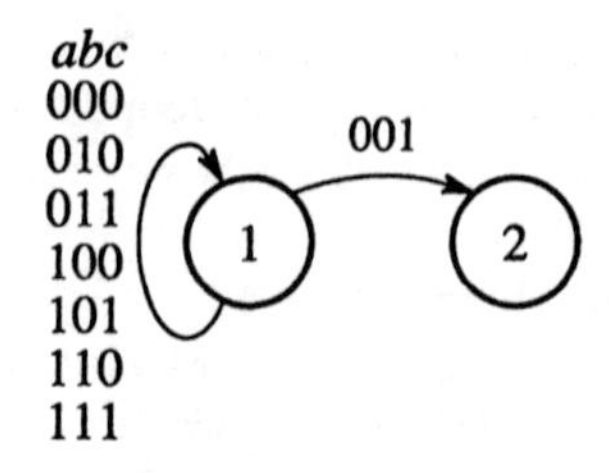

(a) Binary notation

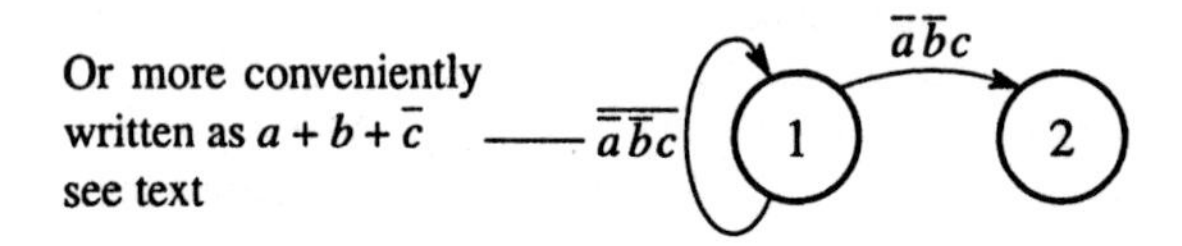

(b) State diagram using transition expressions

Figure 5.15 State diagram notations

values must not make more than one transition expression true as this would create an ambiguous situation.

A state diagram for a *J–K* flip-flop using transition expressions is shown in Figure 5.16.

SELF-ASSESSMENT QUESTION 5.7

Draw the state diagram for a *D*-type flip-flop using transition expressions.

Design problem – pulse generator with variable pulse width

To take an example of using state diagrams with transition expressions, consider the following problem. A synchronous sequential logic circuit is to be designed with two inputs, x_2 and x_1, and a single output, *Z*. The output is to be a 1 for a number of clock cycles given by the two inputs. For example, if $x_2x_1 = 01$, $Z = 1$ for one clock cycle immediately following the input change. If $x_2x_1 = 10$, $Z = 1$ for two clock cycles. If $x_2x_1 = 11$, $Z = 1$ for three clock cycles. If the input changes during the period that the

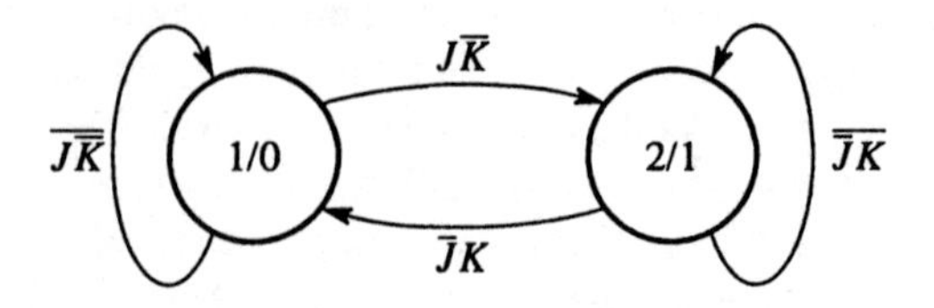

Figure 5.16 Moore model state diagram of the J–K flip-flop using transition expressions

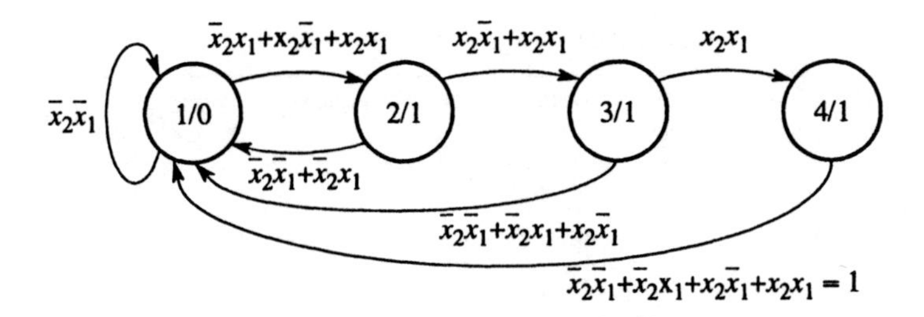

Figure 5.17 State diagram with transition expressions

output is a 1, the output immediately returns to a 0. This problem, a form of pulse generator, naturally suits a Moore model state diagram, as we can define states in which the output is to be 1.

Four states are needed:

State 1 – an initial state
State 2 – for the first clock period that the output is a 1,
State 3 – for the second clock period that the output is a 1, and
State 4 – for the third clock period that the output is a 1.

The state diagram using transition expressions is shown in Figure 5.17. Notice that state 4 returns to state 1 irrespective of the input conditions. Every possible combination of x_1x_2 that causes a particular transition is listed as sum-of-product transition expressions, which reduces the risk of omitting an input combination. Working through the design, the state table in Table 5.10 is obtained.

Table 5.10 *State table for pulse generator*

Present state	*Next state*				*Output, Z*
	x_2x_1				
	00	*01*	*11*	*10*	
1	1	2	2	2	0
2	1	1	3	3	1
3	1	1	4	1	1
4	1	1	1	1	1

For completeness, let us perform the state minimization procedure. Table 5.11 shows the results of this procedure. Notice that state 1 is output incompatible with the other three states, and no minimization is possible.

Table 5.11 *Equivalent pairs of states from Table 5.10*

Possibly equivalent pairs	*Required equivalent pairs*	*Output incompatible*
2, 3	(1, 3), (3, 4)	x
2, 4	(1, 3)	x
3, 4	(1, 4)	x

Next we make assignment of state variables using two state variables, representing state 1 = 00, state 2 = 01, state 3 = 11 and state 4 = 10. (States 2, 3 and 4 can go to state 1, so rule 1 is applied to states 2, 3 and 4.) The procedure gives us the assigned state table shown in Table 5.12.

Table 5.12 *Assigned state table for pulse generator*

Present state	*Next state, Y_2Y_1*				*Output, Z*
	x_2x_1				
y_2y_1	*00*	*01*	*11*	*10*	
00	00	01	01	01	0
01	00	00	11	11	1
11	00	00	10	00	1
10	00	00	00	00	1

Next the flip-flop input functions and output functions are derived, as shown in Figure 5.18, assuming *D*-type flip-flop implementation, leading to:

$$Y_2 = \overline{y_2}y_1x_2 + y_1x_2x_1$$
$$Y_1 = \overline{y_2}\,\overline{y_1}x_1 + \overline{y_2}x_2$$
$$Z = \overline{\overline{x_2}\,\overline{x_1}}$$

We choose to take the inverse function from the 0's on the Karnaugh map for the output function; a single NAND gate can then be used to obtain the function. The final circuit is shown in Figure 5.19.

State table with transition expressions

When using state transition expressions, it may be useful to use a state table which specifically lists the transition expressions, as shown in Table 5.13 for our pulse generator problem. The transition expressions are placed vertically in this table.

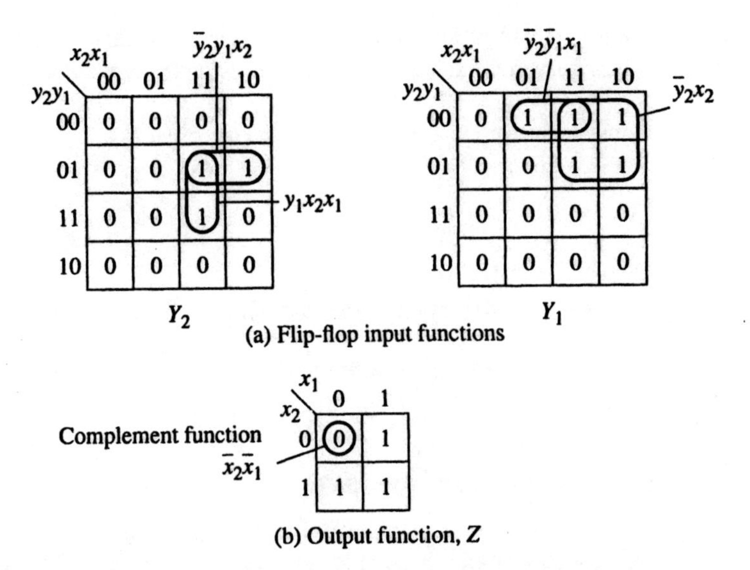

Figure 5.18 Pulse generator flip-flop input and circuit output functions

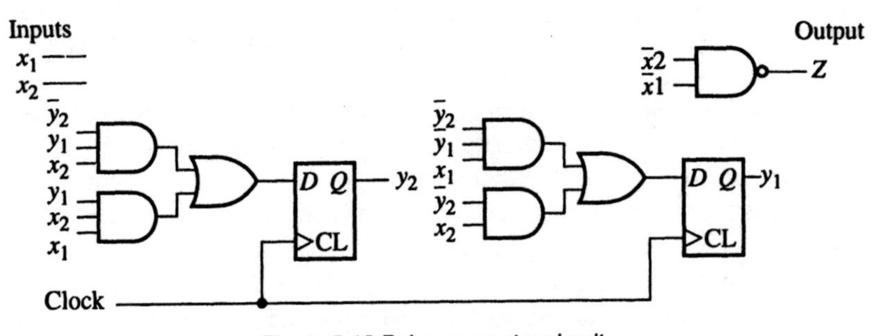

Figure 5.19 Pulse generator circuit

Table 5.13 *State table listing state transition expressions*

Present state	y_2y_1	*Transition expressions*	*Next state*	Y_2Y_1	*Output, Z*
1	0 0	$\bar{x}_2\bar{x}_1$	1	0 0	0
		$\bar{x}_2x_1 + x_2\bar{x}_1 + x_2x_1$	2	0 1	
2	0 1	$\bar{x}_2\bar{x}_1 + \bar{x}_2x_1$	1	0 0	1
		$x_2\bar{x}_1 + x_2x_1$	3	1 1	
3	1 1	$\bar{x}_2\bar{x}_1 + \bar{x}_2x_1 + x_2\bar{x}_1$	1	0 0	1
		x_2x_1	4	1 0	
4	1 0	1	1	0 0	1

From the state table, we can write down "unminimized" next state equations by looking at the entries for which each Y variable is a 1, using the equation:

$$Y_i = \sum\{(\text{present state leading to } Y_i = 1) \cdot (\text{transition expression})\}$$

i.e.:

$$Y_2 = (\bar{y}_2 y_1)(x_2\bar{x}_1 + x_2x_1) + (y_2y_1)(x_2x_1)$$
$$Y_1 = (\bar{y}_2\bar{y}_1)(\bar{x}_2x_1 + x_2\bar{x}_1 + x_2x_1) + (\bar{y}_2y_1)(x_2\bar{x}_1 + x_2x_1)$$

which become the flip-flop input expressions for D-type flip-flops. The expressions simplify to the previous results.

5.2.7 Incompletely specified systems

We have seen the concept of an incompletely specified system in combinational logic circuits in Chapter 3 (Section 3.4.2). In such systems, *don't cares* are used to indicate that certain values are not defined and each could be assumed to take on a 0 or a 1 for minimization purposes.

The design process for incompletely specified systems is as before, except now the state minimization process can consider the don't cares as having values to reduce the number of states, in a similar way as don't cares on a Karnaugh map can help obtain a simpler solution. When don't cares exist, the state table will have don't care entries, marked with X's. Two states can be combined if output compatible, and required equivalent pairs are output compatible. Where an X appears, it may be considered as a valid state number and combined with others. However, as in a Karnaugh map, an X can only be assumed to have one value. If we are using state transition expressions, all the expressions as a set around a state do not need to cover all combinations of variables. Some combination of variables may be don't cares.

Design problem – interface logic

Consider the following computer interface problem. A circuit is to be designed which accepts two inputs called $\overline{RD}$ (for read) and $\overline{WR}$ (for write), and produce a logic output pulse (Z) of duration of two clock periods if $\overline{RD} = 0$ or duration of three clock periods if $\overline{WR} = 0$. We are told that $\overline{RD}$ and $\overline{WR}$ cannot be both a 0 together (i.e. read and write cannot occur simultaneously). Hence $\overline{RD} = 0$, $\overline{WR} = 0$ becomes a don't care condition. Note, in this problem, that the inputs are active low signals, that is a 0 cause action rather than a 1. (Control signals are often active low in computer interface systems.) With this notation, the true variables RD and WR will still cause action when a 1.

The design problem is similar to the pulse generator problem in the last section, but differs in its assumptions. We shall assume that once $\overline{RD}$ or $\overline{WR}$ has activated a pulse, the signal may return to the inactive level (a 1) before the pulse has completed,

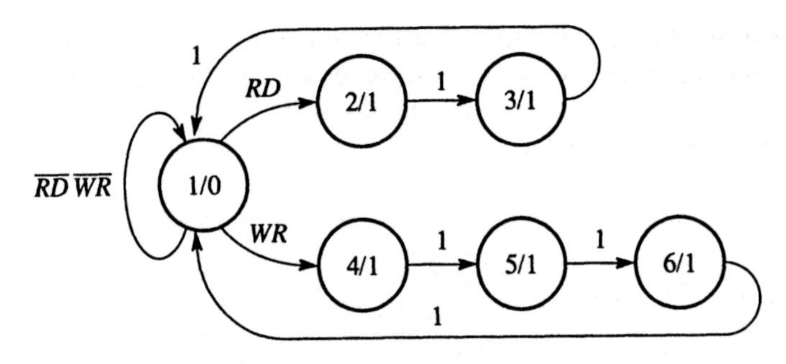

Figure 5.20 State diagram for interface problem using state transition expressions

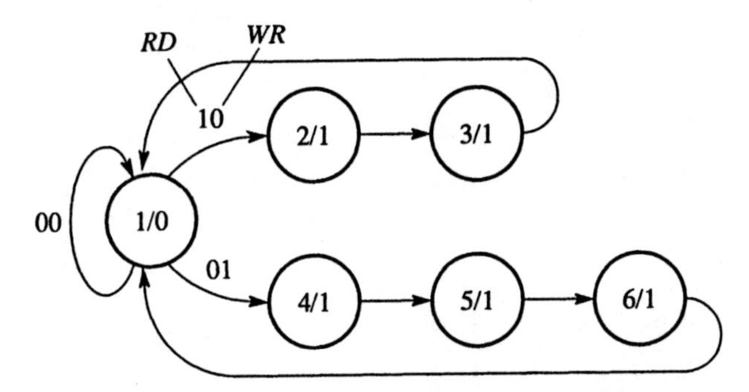

Figure 5.21 State diagram for interface problem

i.e. the value of $\overline{RD}$ or $\overline{WR}$ is a don't care once the pulse has started. Let us first use a Moore model state diagram, and also an intuitive non-minimal state diagram as shown Figure 5.20 to show the usefulness of state minimization. This figure uses state transition expressions. Figure 5.21 shows the same state diagram using Boolean values. For state minimization, it will be more convenient to use the Boolean version.

The state table is shown in Table 5.14. Working through incompatible pairs, we

Table 5.14 *State table for interface problem*

Present state	*Next state*				*Output, Z*
	RD, WR				
	00	*01*	*11*	*10*	
1	1	4	X	2	0
2	3	3	X	3	1
3	1	1	X	1	1
4	5	5	X	5	1
5	6	6	X	6	1
6	1	1	X	1	1

get the table shown in Table 5.15. State 1 is output incompatible with all the other states and need not be considered further. We see that only one pair of states is equivalent, namely states 3 and 6 (actually they are the same), which leads to the state diagram shown in Figure 5.22.

Table 5.15 *Equivalent pairs of states from Table 5.14*

Possibly equivalent pairs	*Required equivalent pairs*	*Output incompatible*
2, 3	(1, 3)	x
2, 4	(3, 5)	x
2, 5	(3, 6)	x
2, 6	(1, 3)	x
3, 4	(1, 5)	x
3, 5	(1, 6)	x
3, 6		
4, 5	(5, 6)	x
4, 6	(1, 5)	x
5, 6	(1, 6)	x

We could have formulated the problem and state diagram differently. For example the state diagram in Figure 5.23 will satisfy the problem specification. We have used the transition expression $RD \oplus WR$ to signify that the transition occurs if $RD = 1$ or $WR = 1$ but not both (i.e. exclusive-OR). Also, input values are specified for subsequent transitions – if these values are not maintained, the pulse is terminated early as in the pulse generator problem in the previous section. However, not all possible combinations are included in Figure 5.23 as we assume certain combinations cannot occur, notably $RD = WR = 1$. In this state diagram, we also show the transition

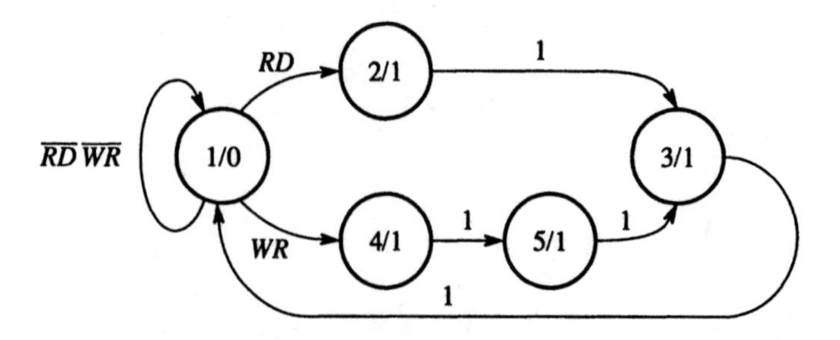

Figure 5.22 Reduced state diagram for interface problem

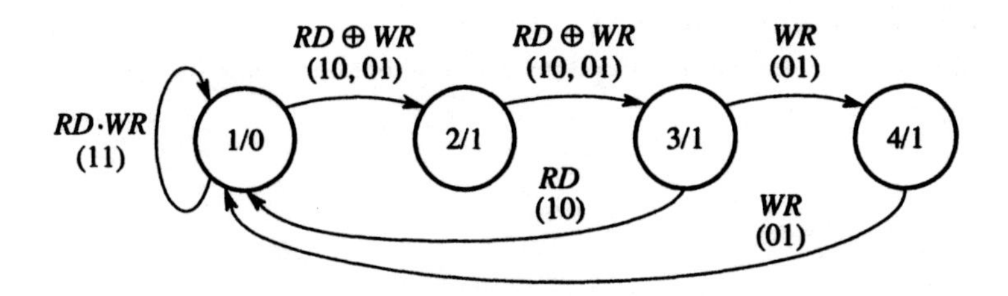

Figure 5.23 Alternative state diagram for interface problem (different assumptions)

variable values in parentheses. It is left as an exercise to determine whether this particular state diagram minimizes.

In a system having don't cares, if state 1 ≡ state 2 and state 2 ≡ state 3, then state 1 ≡ state 3 only if the don't cares are interpreted in the same way for each equivalence. This would occur if we find state 1 ≡ state 2, state 2 ≡ state 3 and state 1 ≡ state 3 (i.e. a completely circular equivalence).

5.2.8 One-hot assignment

In all the designs so far, we have attempted to reduce the number of flip-flops to a minimum by encoding the states. An alternative is to not encode the states, but use one state variable for each state. For example, if there are eight states, eight state variables would be used, say $y_0y_1y_2y_3y_4y_5y_6y_7$. One of these variables is a 1 for each state. So y_0 could be for state 1, y_1 for state 2 and so on. Essentially the state assignment would be as follows:

	y_0	y_1	y_2	y_3	y_4	y_5	y_6	y_7
State 0 =	1	0	0	0	0	0	0	0
State 1 =	0	1	0	0	0	0	0	0
State 2 =	0	0	1	0	0	0	0	0
State 3 =	0	0	0	1	0	0	0	0
State 4 =	0	0	0	0	1	0	0	0
State 5 =	0	0	0	0	0	1	0	0
State 6 =	0	0	0	0	0	0	1	0
State 7 =	0	0	0	0	0	0	0	1

This form of assignment is known as *one-hot assignment.* Though one-hot assignments leads to more state flip-flops – eight flip-flops in the above example – the next state functions will generally be simpler. A common variation of one-hot assignment is an *almost* one-hot assignment which reserves the pattern 00... 00 for the initial state. It is then easy to initialize the state flip-flops to the initial state. For eight states, seven state variables would be needed.

SELF-ASSESSMENT QUESTION 5.8

List the state assignment for a system with eight states using almost one-hot assignment.

5.3 Summary

This chapter has covered the following:

- General synchronous sequential circuit models (Mealy model and Moore model).
- Design procedures for designing synchronous sequential circuits.
- Detailed methods of obtaining efficient designs.
- Methods of reducing the component count using state minimization.
- A form of state diagram using transition expressions which is suitable for more complex sequential circuits.

5.4 Tutorial questions

5.1 A synchronous sequential circuit has one input, x, and one output, Z. The output is to become a 1 whenever the input sequence received is 0110100, otherwise the output is to be 0. Draw the Moore model state diagram for the circuit.

5.2 Draw the Mealy model state diagram for Question 5.1.

5.3 A synchronous sequential logic circuit has two inputs, x_1 and x_2, and one output, Z. The output is required to become a 1 only during the presence of the final number in the sequence 00, 11, 00, 11 applied to the two inputs. Draw the Moore model state diagram for the circuit and derive the state table. Hence design a circuit to perform the desired function using the minimum number of D-type flip-flops.

5.4 Repeat Question 5.3 using J–K flip-flops.

5.5 A synchronous sequential circuit has one input, x, and one output, Z. The output is to become a 1 whenever the input sequence received is 1100 or 0011, otherwise the output is to be 0. Draw the Moore model state diagram for the circuit and derive the state table. Hence design a circuit to perform the desired function using the minimum number of D-type flip-flops.

5.6 Repeat Question 5.5 using a Mealy model state diagram.

5.7 Design a pulse generator which generates a pulse width of n clock periods, where n is input to the circuit on three lines ($0 < n \leq 7$)

5.8 Minimize the Moore model state table given in Table 5.16.

Table 5.16 *State table for Question 5.8*

Present state	*Next state*				*Output, Z*
	x_1x_2				
	00	*01*	*11*	*10*	
1	2	3	2	1	0
2	1	3	1	4	1
3	3	4	3	5	1
4	5	3	2	2	0
5	6	7	1	4	1
6	3	7	1	1	0
7	1	1	1	2	1

5.9 Minimize the Mealy model state table shown in Table 5.17.

Table 5.17 *State table for Question 5.9*

Present state	*Next state*				*Next output, Z*			
	x_1x_2							
	00	*01*	*11*	*10*	*00*	*01*	*11*	*10*
1	1	2	2	4	0	0	0	0
2	1	3	1	4	0	0	0	0
3	1	3	2	4	0	0	0	0
4	1	8	7	5	0	0	0	0
5	1	5	6	5	0	0	1	0
6	3	8	7	5	0	0	0	0
7	2	8	6	5	0	0	0	0
8	3	8	7	8	0	0	1	0

5.5 Suggested further reading

Katz, R. H., *Contemporary Logic Design*, Benjamin/Cummings: Redwood City, California, 1994.

Nelson, V. P., Nagle, H. T., Carroll, B. D. and Irwin, J. D., *Digital Logic Circuit Analysis and Design*, Prentice Hall: Englewood Cliffs, New Jersey, 1995.

Wakerly, J. F., *Digital Design Principles and Practices, 2nd edition*, Prentice Hall: Englewood Cliffs, New Jersey, 1994.

We have omitted any consideration for asynchronous sequential circuit design, and most other books limit their exposition in this area. A treatment of asynchronous sequential circuit design can be found in Wakerly (1994).

CHAPTER 6

Designing with programmable logic devices

Aims and objectives

In the previous chapters, we described the fundamental logic components that make up a digital system: gates, combinational circuits formed with gates such as decoders, flip-flops and sequential circuits designed with flip-flops and gates. In this chapter, we shall explore a flexible logic component called a programmable logic device, PLD. A PLD has an internal structure which can be configured as various combinational circuits and often sequential circuits. PLDs are used to reduce the component count of complex logic systems substantially.

6.1 Programmable logic devices (PLDs)

Programmable logic devices contain gates and in some cases flip-flops arranged so that the interconnections between the components can be altered to implement various logic functions. Combinational logic PLDs, those containing only basic logic gates, are usually organized as an array of AND gates and OR gates that implement sum-of-product expressions. Sequential PLDs add flips-flops to the outputs.

Whatever logic arrangement, a general requirement of a PLD is to have a means of changing the interconnections to form a different logic configuration. The original permanent but selectable way of achieving this was to manufacture the devices with semiconductor fuses (integrated circuit fuses). Initially these fuses are intact within the device which provides extensive interconnections. Selected fuses are then "blown" by the user to obtain the desired interconnections using special *PLD programmer* units. Commonly, the PLD programmer is attached to a small computer, and software is provided to translate the desired logical functions to the corresponding internal interconnection patterns. Of course, once a fuse is blown, the connection cannot be remade. PLDs that the user can program, such as these semiconductor fuse PLDs, are called "field programmable".

The semiconductor fuse PLDs were developed in the 1970s and now field programmable versions of PLDs exist which use semiconductor erasable read-only memory technology.[1] Here the connections depend upon stored binary information. A memory cell exists at each connection point to store a 0 (say) to maintain a connection or a 1 to disable a connection. Semiconductor memory-based PLDs provide the capability to alter the interconnections rapidly and many times.

Manufacturers also produce versions of PLDs with permanent connections according to a user's specification. To achieve the fixed connections, specific integrated circuit masks must be created to manufacture the device. This process is expensive for manufacturing one device, but cost becomes much less important when manufacturing a large number of identical devices.

All three types of PLDs (using semiconductor fuses, memory technology, or manufactured connections) have produced matching package pin-outs so that one type can be substituted with another type in a system. A re-programmable type might be used initially for experimentation, though these are the slowest in logic operation and the most expensive; then, perhaps, the one-time field programmable devices might be used for a small production run; and finally the manufactured PLDs with specified connections can be used for a large production run.

In the following, we shall discuss PLDs in terms of "programmable connections", which applies to all PLDs whether using semiconductor fuses, memory technology, or manufactured connections. The logic structures of the different implementations are similar. First let us consider combinational circuit PLDs.

6.2 Combinational circuit PLDs

6.2.1 Programmable logic arrays (PLAs)

The usual logic structure for a combinational circuit PLD is one which creates sum-of-product expressions. A convenient interconnection structure is a two-dimensional array with programmable connections at the crossover points. Figure 6.1 shows the logic diagram of a sample combinational circuit PLD with sixteen inputs to the circuit ($I_0, I_1, I_2, I_3, I_4, I_5, I_6, I_7, I_8, I_9, I_{10}, I_{11}, I_{12}, I_{13}, I_{14}, I_{15}$) and eight independent outputs ($F_0, F_1, F_2, F_3, F_4, F_5, F_6, F_7$). Combinational circuit PLDs of this sort are called *programmable logic arrays* (PLAs). The number of inputs and outputs can vary between different PLAs.

Considering Figure 6.1, each output can produce a sum-of-product expression consisting of up to forty-eight product terms, and each product term can have up to sixteen input variables or their inverse in this example. The number of product terms is determined by the number of AND gates, and the number of variables in each product term is determined by the number of inputs to the circuit. Both true and inverse values of each input are produced which can be selected for the inputs to the AND gates.

The actual internal implementation of the PLA uses only one line where thirty-

---

[1] More details of memory technology can be found in Suggested further reading at the end of the chapter.

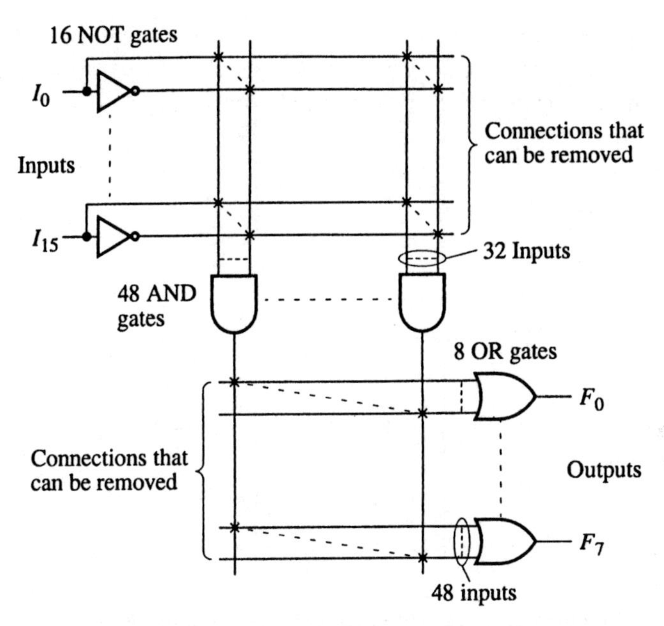

Figure 6.1 Field programmable logic array logic diagram

two are indicated into each AND gate in Figure 6.1, and only one line where forty-eight are indicated into each OR gate in Figure 6.1. The crossover connections are then done with a semiconductor fuse in series with a diode (or equivalent connection) to form the inputs of a gate.[2] Hence we usually draw the PLA as shown in Figure 6.2.

There are four conditions for each input variable as shown in Figure 6.3: the initial unprogrammed state with all connections left intact with both the true and inverse variables connected. Either the true variable or the inverse variable can be selected by removing one connection. If both connections are removed, no variable is selected, i.e. this variable does not occur in the product term. If all the connections to one AND gate are removed, a logic 1 output is often generated.

Both the true and inverse signals of the same variable would not be used together in one product term as this would make the product term a 0 permanently (e.g. I_i and $\bar{I}_i$ are not used together as $PI_i\bar{I}_i = 0$ where P represents the other variables in the product term). An exception to this is dealing with completely unused AND gates, in which case the unprogrammed state would be used, with all connection to that AND gate left intact to make the corresponding product term a 0. (It is theoretically possible for logic glitches to occur if all the variables were to change simultaneously but this situation is very unlikely to occur.) An alternative is to apply a logic 1 to one input which is then connected to each of the unused AND gates. In some of the following, the details for the unused AND gates are omitted for clarity.

---

[2] More details of implementations of gates can be found in Suggested further reading at the end of the chapter.

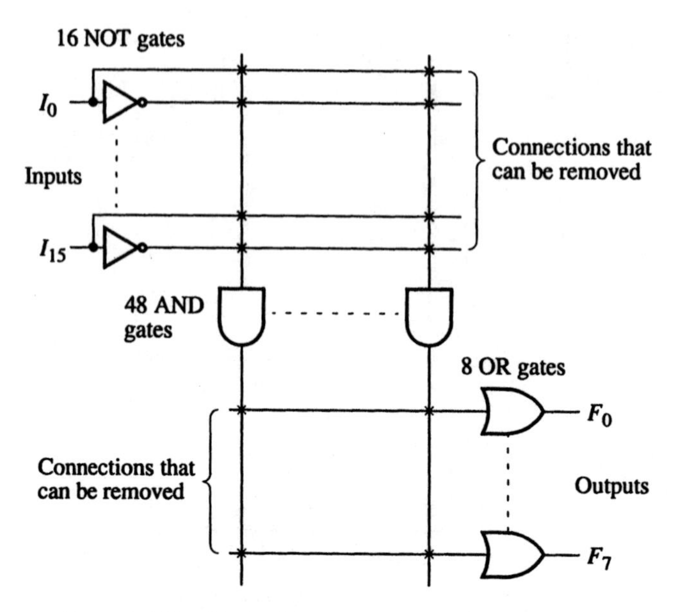

Figure 6.2 Simplified field programmable logic array diagram

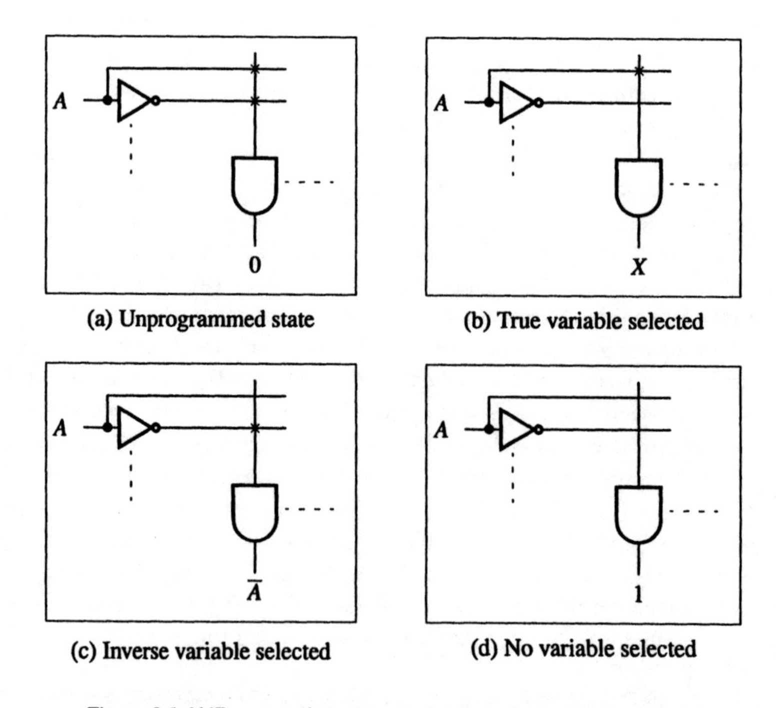

Figure 6.3 AND connection states in field programmable logic array

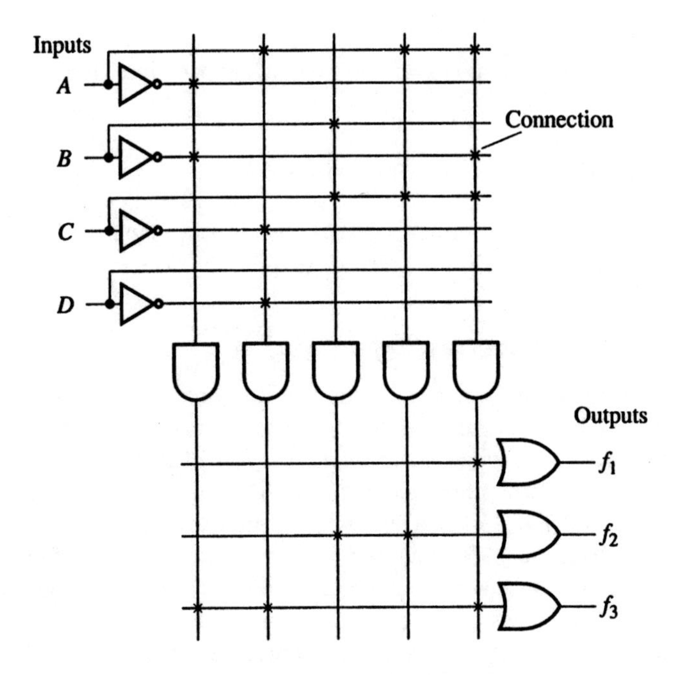

Figure 6.4 Using a PLA to generate the functions $f_1 = A\overline{B}C$, $f_2 = AC + BC$, $f_3 = \overline{A}\,\overline{B} + A\overline{C}\,\overline{D} + A\overline{B}C$

DESIGN EXAMPLE

Suppose the functions:

$$f_1 = A\overline{B}C$$
$$f_2 = AC + BC$$
$$f_3 = \overline{A}\,\overline{B} + A\overline{C}\,\overline{D} + A\overline{B}C$$

are to be implemented with a PLA. The PLA connections for these functions is shown in Figure 6.4. The X's show where the connections have been left intact. Notice that when a common product term exists between functions, this can be shared as in a normal AND–OR implementation. Logic simplification may be useful even with a PLA implementation, because of the limited number of gates. A PLA is actually best used when several sum-of-product expressions are required, especially if the sum-of-product expressions have many more terms than are given in this example.

SELF-ASSESSMENT QUESTION 6.1

Draw a normal logic diagram for Figure 6.4.

If a function to be implemented is not in sum-of-product form, it needs to be converted into this form.

EXAMPLE

To implement the function $f = C(A \oplus B)$, first expand into sum-of-product form:

$$f = C(A \oplus B) = C(A\bar{B} + \bar{A}B) = A\bar{B}C + \bar{A}BC$$

and then implement as normally. (PLD software is available which will convert Boolean expressions into the appropriate form, see later.)

6.2.2 Programmable array logic (PAL)

The PLA described in the previous section has both programmable connections to the inputs to the AND gates and programmable connections to the inputs to the OR gates. Advanced Micro Devices manufacture a family of PLD devices having only programmable connections to the inputs of the AND gates. They call the device a *programmable array logic* (*PAL*), a term which is now widely accepted for this type of device.[3] In a PAL, each AND gate is directly wired to one input of one OR gate. Typically, eight AND gates will be connected to the inputs of each (eight-input) OR gate. One motive for this restriction is to increase the speed of operation as programmable connections incur a greater delay than direct wired connections.

The original PALs, introduced in the 1970s, were designed using bipolar technology with bipolar semiconductor fuses, but now pin-compatible CMOS electrically erasable versions are available. After a rather slow start, PALs have now become very widely used to replace traditional SSI/MSI logic components in applications such as computer interface circuits. The devices now have multiple sources (including Texas Instruments and Signetics Corporation). Sequential logic circuit versions exist (registered PALs, see Section 6.3.1).

An example of a combinational circuit PAL is shown in Figure 6.5, an AMD PAL16L8. The letter L in the device identification indicates that the outputs are inverted (low). In this device, there is a maximum of sixteen inputs and eight outputs, but some outputs are shared with inputs because of pin limitations of the package.[4] The input/output connections do provide convenient feedback paths. The outputs of the device are provided with three-state buffers which can be enabled by a product function of the inputs and feedback variables. The shared input/output connections can be used as inputs when the outputs are disabled.

In Figure 6.5, there are ten independent inputs, $I_1, I_2, I_3, I_4, I_5, I_6, I_7, I_8, I_9$ and I_{10}, each with both true and inverse selection. There are eight outputs, six of which also serve as inputs, $IO_2, IO_3, IO_4, IO_5, IO_6, IO_7$, and two are dedicated outputs, O_1 and O_8. This device can create eight sum-of-product expressions, each having seven product terms and each product term having ten variables. Alternatively two independent sum-of-product terms could be created each having seven product terms

---

[3] PAL is a registered trademark of Advanced Micro Devices Inc.

[4] PALs are often packaged in 20-pin or 24-pin dual-in-line packages; the PAL16L8 is in a 20-pin package.

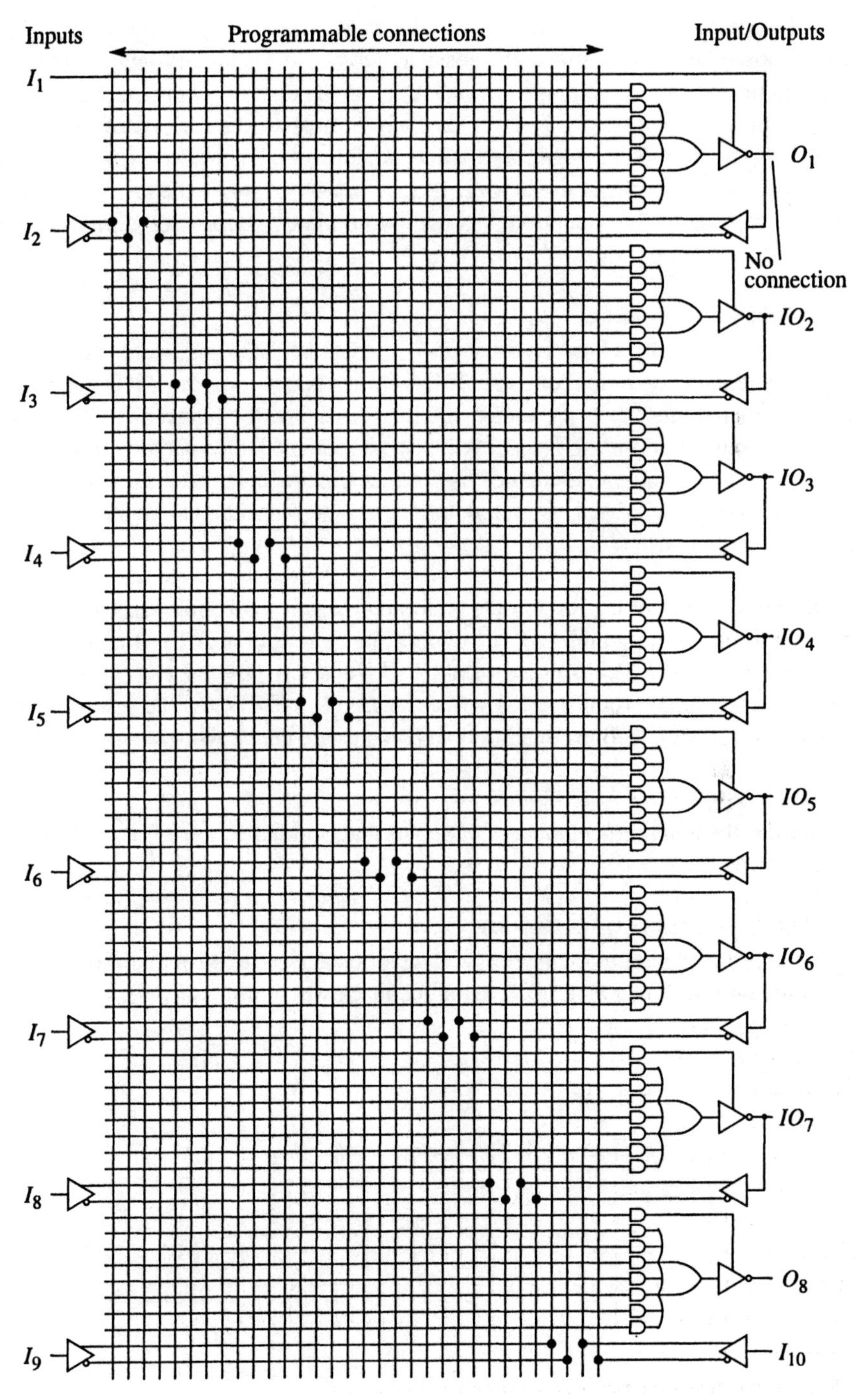

Figure 6.5 Programmable array logic (AMD PAL16L8)

each having up to sixteen variables (six being from the shared input/output pins). Numerous other combinations are possible subject to the shared input/output line limitations. We do not gain any advantage in having shared product terms between sum-of-product expressions in PALs as each term has to be created separately for each OR gate. In the PAL16L8, the outputs are inverted, i.e. the functions are AND–OR–NOT. Larger combinational logic PALs exist, for example the PAL20L8 which has fourteen inputs, six input/outputs and two outputs, i.e. providing up to twenty inputs in a 24-pin package.

DESIGN EXAMPLE

Figure 6.6 shows how the function $\bar{f} = \bar{A}BC + A\bar{B}C + \bar{A}\,\bar{C}$ would be implemented on one output of a PAL16L8. Notice that the unused AND gates have all their connections left intact which forces a 0 on the inputs of these gates. The outputs of the AND gates will be a permanent 0, thus not affecting the OR gate. The output, $\bar{f}$, is enabled by removing all the connections on the input of the AND gate driving the output buffer, as a permanent 1 is required to enable the output buffer.

A sum-of-product expression can be implemented with more than two levels, perhaps to fit into a specific PAL, by connecting an output back to an input, such as shown in Figure 6.7. For example, to obtain $A + B + C + D + E + F + G + H + I + J$, the first stage could produce $X = A + B + C + D + E + F + G$ and the second stage = $X + H + I + J$. Connections to unused AND gates are not shown here for clarity.

SELF-ASSESSMENT QUESTION 6.2

What is the maximum number of variables that could be in a product term to fit in a PAL16L8?

SELF-ASSESSMENT QUESTION 6.3

What is the maximum number of product terms that could be in a sum-of-product expression to fit in a PAL16L8, using feedback where necessary?

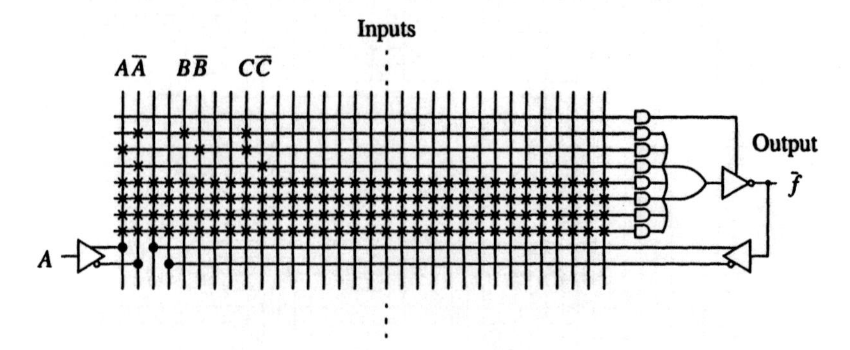

Figure 6.6 Implementing the function, $\bar{f} = \bar{A}BC + A\bar{B}C + \bar{A}\bar{C}$ using a PAL

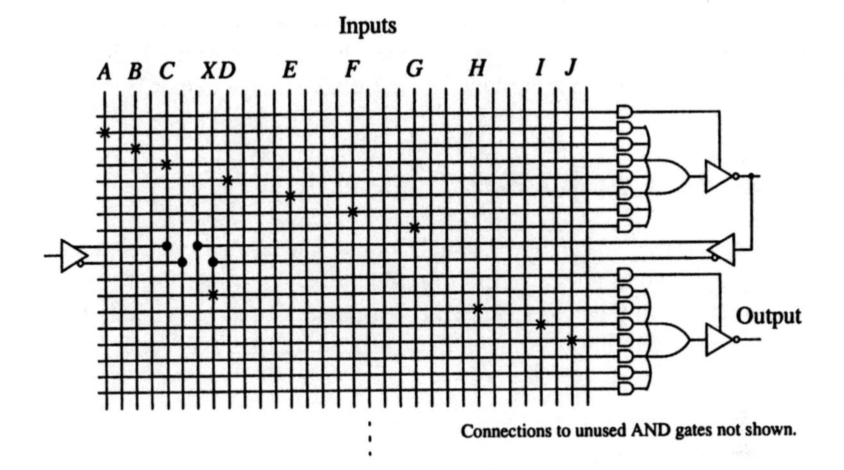

Figure 6.7 Multilevel sum-of-product expression

The AMD PAL16L8 described in Figure 6.5 has inverting buffers, creating active-low outputs (the letter L in the name indicating inverting outputs). PALs with non-inverting outputs do exist (with the letter H in the name indicating non-inverting outputs). Actually if you have a device which naturally creates inverting outputs, and true outputs are required, the function can be simply manipulated into the inverse function, though the number of product terms created may exceed that possible for a particular PAL implementation. The easiest way of deducing the inverse function as a sum-of-product expression is to group the 0's on the Karnaugh map of the function rather than the 1's.

In some devices, either active-low or active-high outputs can be programmed by incorporating a programmable true/inverse circuit, such as that shown in Figure 6.8. The programmable connection left intact will create a true output (i.e. Output = Input) while removing the programmable connection will create an inverse output (i.e. Output = $\overline{\text{Input}}$). The circuit in Figure 6.8 would typically be placed after the OR gate and before the output buffer. These PALs are sometimes called *generic array logic devices* (GALs[5]) and come under the category of *universal* PALs. An

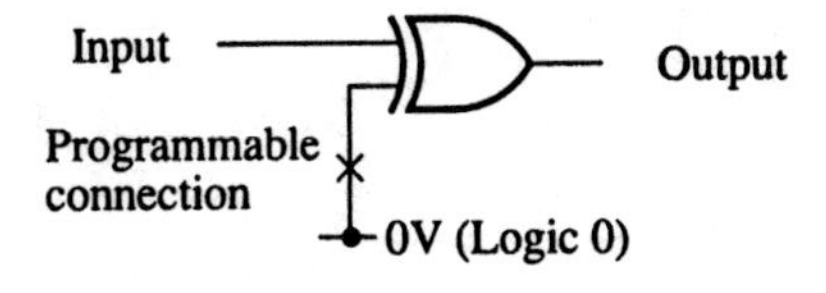

Figure 6.8 True/inverse output circuit

[5] GAL is a trademark of Lattice Semiconductor Corp.

example of a universal PAL programmable for either true or inverse outputs is the AMD PALCE16V8 which has the same AND–OR structure as the PAL16L8. (The C stands for CMOS, the E for erasable, V for varied/variable output.)

6.3 Sequential circuit PLDs

PLA/PALs can be used to create sequential circuits. The feedback connections could, theoretically, be used to produce flip-flops. More likely, separate flip-flops could be employed as shown in Figure 6.9. Since this is a general sequential circuit, it is attractive to incorporate the flip-flops into a PLD, and such PLDs exist, called *registered PLDs*.

6.3.1 Registered PLDs

In registered PLDs, flip-flops consisting of *D*-type, or *S–R*, or *J–K* flip-flops are integrated into the device. The term *registered* implies that all the outputs will be clocked by a single clock signal and, hence, in the device the clock input of each flip-flop is connected to a common clock line.

A representative registered PAL design is the AMD PAL16R6 shown in Figure

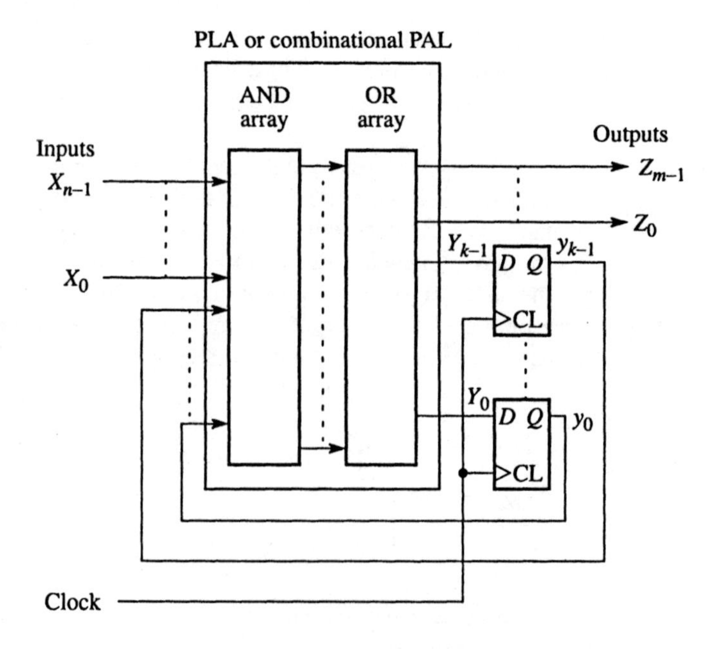

Figure 6.9 Synchronous sequential circuit using PLA/PAL

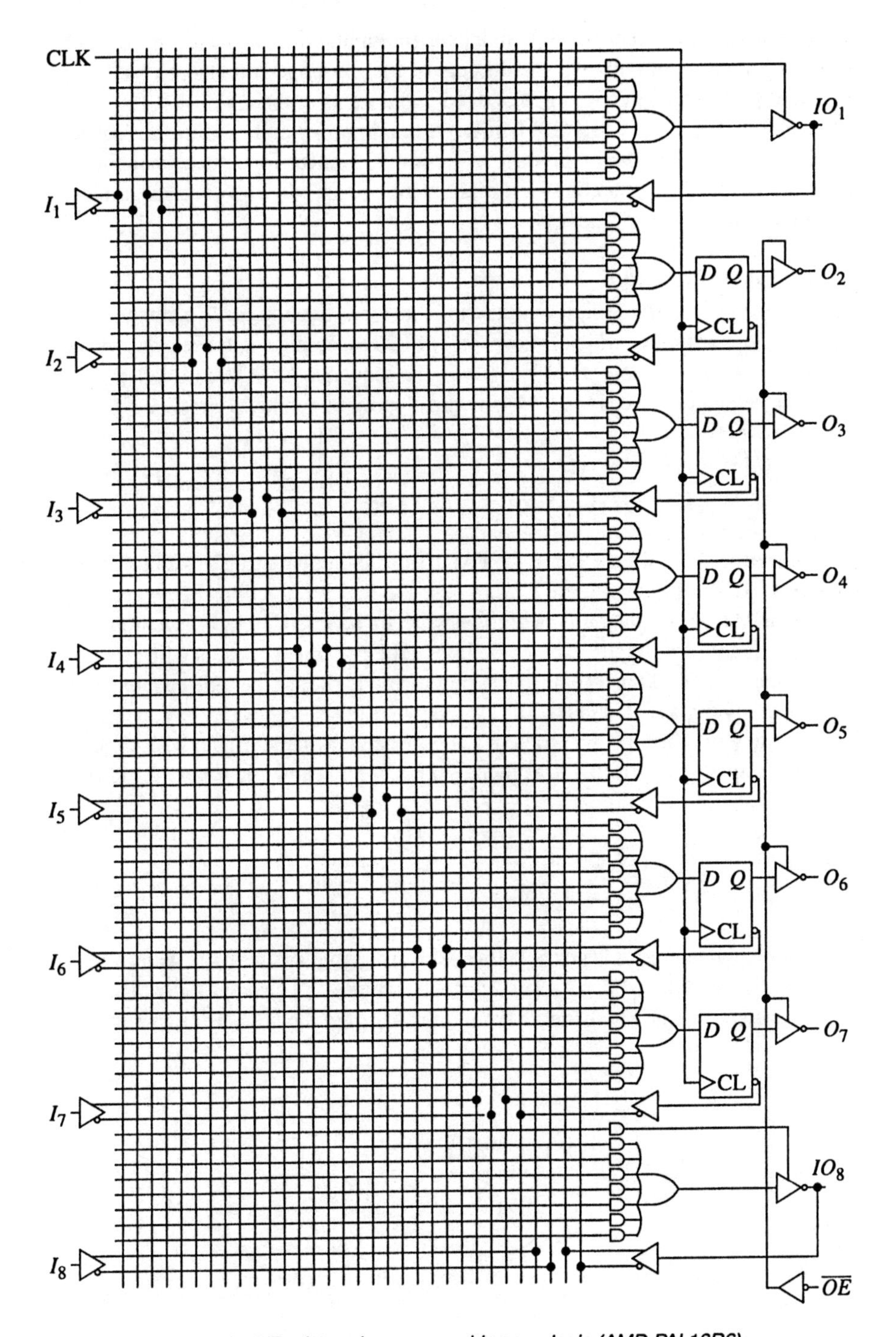

Figure 6.10 Registered programmable array logic (AMD PAL16R6)

6.10. In this device, there are eight independent inputs, six D-type flip-flops and two PAL16L8 type input/outputs. The outputs of the flip-flops can also be fed back as variables providing a maximum of sixteen variables as in the PAL16L8. (In fact, the PAL16L8 is part of the PAL16R8 family.) A maximum of eight product terms can be implemented in the sum-of product expressions. Hence, for a state machine, this device provides for a maximum of eight x inputs, six y/Y state variables and two separate sum-of-product Z outputs.

DESIGN EXAMPLE

Consider the first design example in Chapter 5, Section 5.2.1, namely a synchronous counter which follows one of two repeating sequences, dependent upon a control input, C. When $C = 0$, the sequence is 00, 01, 11. When $C = 1$, the sequence is 00, 11, 01. D-type flip-flop input equations are:

$$D_2 = \bar{y}_2 y_1 \bar{C} + \bar{y}_2 \bar{y}_1 C$$
$$D_1 = \bar{y}_2 \bar{C} + \bar{y}_2 \bar{y}_1 C + y_2 y_1 C$$

Let us assume we are to use the PAL16R8. Though the true functions do enter the flip-flops in a PAL16R8 (as opposed to inverse functions being generated in a PAL16L8), the outputs of the device are inverted. Hence if we implement the equations directly, we get the bits in the generated sequence inverted, i.e. either 11, 10, 00 or 11, 00, 10. (Coincidentally these repeating sequences are actually 00, 11, 10 and 00, 10, 11, the two original sequences swapped.) If we again work through the problem as described in Chapter 5, but with the bits of the sequence inverted, i.e. as in the state diagram shown in Figure 6.11, we get the functions:

$$D_2 = y_1 \bar{C} + y_2 \bar{y}_1 C + \bar{y}_2$$
$$D_1 = \bar{y}_2 \bar{C} + \bar{y}_2 y_1 + y_2 \bar{y}_1 C$$

For convenience, it is possible to label Karnaugh maps with the pattern 11, 10, 00, 01 rather than 00, 01, 11, 10 since the pattern still has the "Gray code" characteristic of one bit changing from one pattern to the next. Though done here,

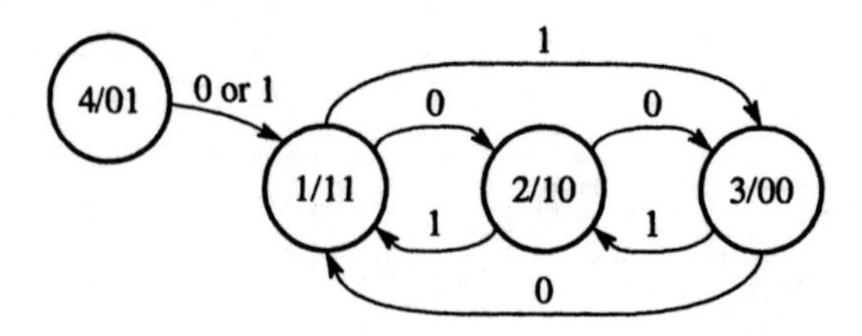

Figure 6.11 Moore model state diagram for counter problem with bits inverted

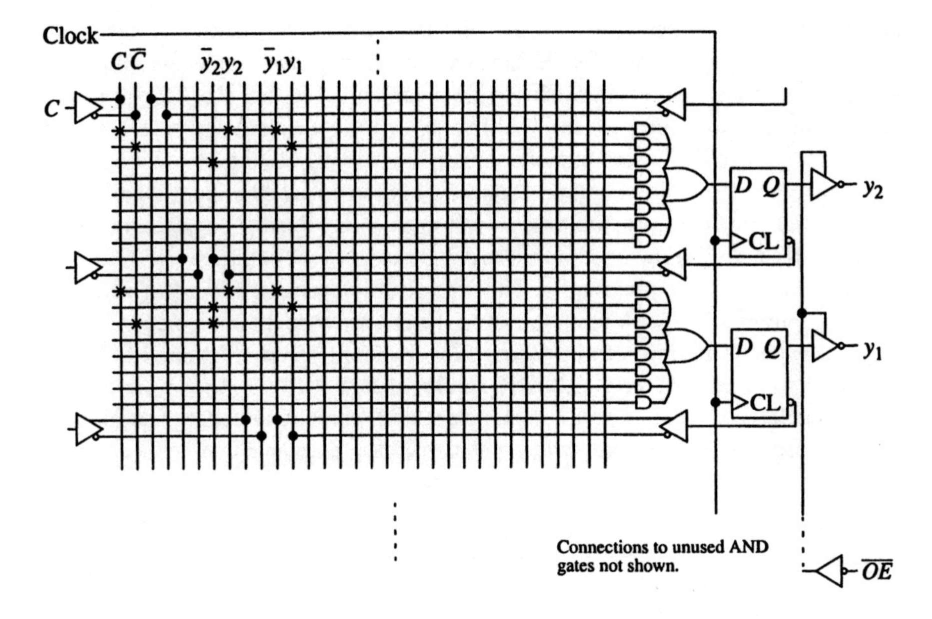

Figure 6.12 Counter using registered programmable array logic

it is not absolutely necessary to minimize the functions for the PAL. The registered PAL implementation is shown in Figure 6.12. Notice how it is not necessary to use input pins for the feedback connections as internal feedback paths exist for y_2 and y_1.

Other AMD registered PAL devices include the AMD PAL20X10 intended for counter designs. This device has AND–OR–exclusive-OR gate combinations as shown in Figure 6.13 and would allow direct implementation of state variable functions consisting of:

(sum-of-product expression) ⊕ (sum-of-product expression)

where each sum-of-product expression can have two terms and each term can have up to sixteen variables (subject to shared input/output pins limitations, as in the PAL16R6).

PLAs with output flip-flops also exist, that is, AND–OR gates with selectable connections to both the AND and OR gates, and "registered" outputs (flip-flops with a common clock). Such devices are called *programmable logic sequencers* (PLSs).

6.3.2 Macrocells

Macrocells are repeated circuits inside a PLD with selectable functions. An example of a macrocell which can select either a flip-flop output or a path around it is shown in Figure 6.14. Now the flip-flop becomes physically further from its output pin and integrated into the macrocell. Such flip-flops are called *buried flip-flops* in macrocells. In this example, there are two multiplexers, each controlled by a programmable connection. When the programmable connection is removed, the input labelled 1 is selected. When the programmable connection is left intact, the input labelled 0 is selected. In Figure 6.14, one multiplexer can select either the flip-flop Q output or the combinational AND–OR function as the device output – in both cases with an extra inversion through the output buffer. The other multiplexer can select either the flip-flop Q output as a feedback input to the AND–OR array or the device output as the feedback input. Variations of this design exist in complex PALs. Examples of PALs with macrocells include the PAL22V10 and PAL32VX10.

Macrocells can also describe other repeated combinational/sequential logic circuits arranged in arrays with two-dimensional programmable interconnects (*logic*

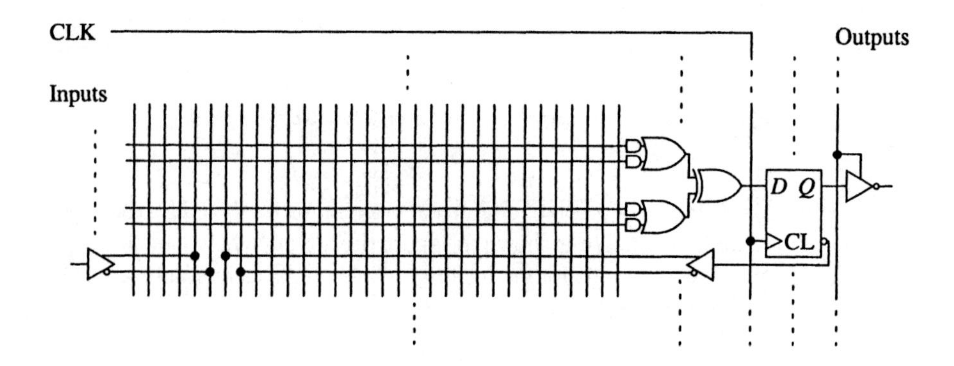

Figure 6.13 AND–OR–exclusive-OR gate combinations in AMD PAL20X10

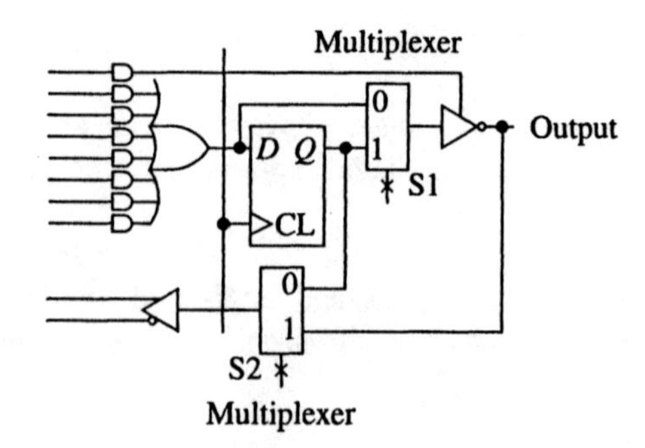

Figure 6.14 Example macrocell with buried flip-flops

cell arrays). The term macrocell is used with a different meaning in VLSI design for predesigned library cells.

6.4 PLD programming tools

To use PLDs, connections to be removed must be identified. For simple arrangements, we could work through the connection patterns from the logic diagram of the PLD, but software tools are normally used to perform this task. This software will take, as input, a specification of the logic function to be realized. The specification can be in the form of Boolean expressions, truth table description, or (for sequential circuits) a state diagram description. The output of the software will include a file holding the required connection patterns (*programming map*). The output file is then used to drive PLD programmer units directly. Some software also includes a logic simulator so that the logic design can be verified before programming the PLDs. Popular software tools are PALASM2 (PAL assembler) from AMD Inc. and ABEL (Advanced Boolean Expression Language)[6] from Data I/O Corporation. Let us concentrate upon ABEL as this software enjoys widespread acceptance.

6.4.1 ABEL source file structure

The user writes a "computer" program following the rules of the ABEL language to specify the logic function to be implemented by the PLD. Statements can specify the device to be used and the names can be allocated to the pins. The output functions are typically described by Boolean expressions but there are other ways.

The program is divided into parts, typically as shown in Figure 6.15. ABEL keywords are not case sensitive, but user defined names are case sensitive. ABEL allows free format using white spaces and newlines. The program must start with `module` and a name, and end with `end` (which must have the same name attached only for multi-module source files). The `title` statement simply provides a header to generated files, as given by a character string enclosed in single quotes following the title keyword. Various declarations follow next. The `device` declaration identifies the device to be programmed. The device specified must be supported by the compiler, and names defined by ABEL must be used.

"Directives" may appear to alter the way the compiler interprets or manipulates the source file. The `@ALTERNATE` directive allows an alternative set of symbols to be recognized (see later).

The signal/pin assignment declaration given with the `pin` keyword associates identifiers with actual device pins. The device pins may be inputs or outputs and the identifiers will be used in the subsequent equation section to represent the signals on the inputs and outputs. Device and pin declaration statements are terminated with a semicolon.

---

[6] ABEL is also a trademark of Data I/O Corporation. Full details of the language can be found in Data I/O Corp. (1990).

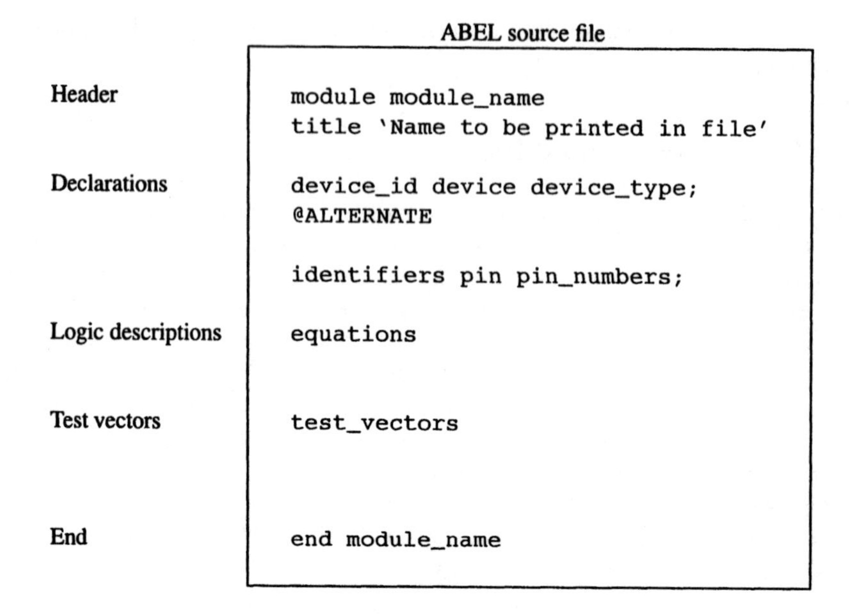

Figure 6.15 Basic ABEL source structure

Next is the logic description section. If the Boolean equation method is used, the `equation` section consists of a list of Boolean expressions defining outputs in terms of the inputs. Each expression is terminated with a semicolon. Alternative ways of describing required logic functions are by giving a truth table, or a text version of a state diagram if a sequential circuit. Comments can be inserted into the source program by enclosing the comment with double quotation marks (or start the comment with double quotes and terminate with a new line).

The optional test vector section defines collection of input values and expected output values. The test vector section is used with the ABEL simulator to establish whether the design is logically correct.

The ABEL *compiler* takes the source file in the form of Figure 6.15 and creates the necessary documentation and connection patterns to program the PLD. The pattern of programmable connections is held in a file and this file is downloaded into the PLD. The documentation lists the final Boolean equations for the PLD, which may have been reduced or rearranged from the original equations to fit into the PLD, and is given in another file.

Architecture independent design

It is not actually necessary to specify the device and pin numbers; this information can be omitted so as to create an *architecture independent design*. Device selection can be suggested by the ABEL software at the time that a device is to be programmed. There may be several choices for the device. Sometimes when the device is not specified initially, the source program must also include additional details of the signals using signal *attribute* assignments (`istype` statements) and *dot extensions* describing features of the signals. For example a *D*-type flip-flop output, Q_0, is defined by `Q0 istype 'reg_d'`. An extension can be attached to a signal name, for example to specify that an input will operate as a flip-flop clear signal (`.RE`). In our subsequent examples, we assume device selection is made early in the design and specified.

6.4.2 Combinational functions

A combinational function consists of input variables connected with logical operators such as AND, OR and NOT, and assigned to an output variable with the equality symbol. The default symbols for AND, OR, NOT, exclusive-OR and exclusive-NOR are `&`, `#`, `!`, `$` and `!$` respectively. These are not very clear so an alternative set is also available, `*` (for AND), `+` (for OR), `/` (for NOT), `:+:` (for exclusive-OR) and `:*:` (for exclusive-NOR). The alternative set is chosen by using the `@ALTERNATE` directive. Notice that the alternative set of symbols precludes the use of arithmetic addition, multiplication, and divisor operators which are also available in ABEL language. To describe a combinational function such as:

$$f = A\overline{B} + CD$$

the function in ABEL is written as:

```
f = A * /B + C * D;
```

or with the default symbols:

```
f = A & !B # C & D;
```

The order of execution of operators is defined in a priority list from 1 to 4 (see Data I/O Corp., 1990) and can be overridden by the use of parentheses. For our simple equation, this is unnecessary as NOT has the highest priority, then AND, and then OR, and the expression will compute as expected. The output f and inputs A, B, C and D have to be declared in the declaration section, and associated with specific device pins (unless an architecture independent design).

EXAMPLE

One application of a PLD is to implement a decoder logic circuit, an application which occurs frequently in computer design. Suppose a decoder is to have four inputs, A, B, C and D, and four outputs, Q_0, Q_1, Q_2 and Q_3. Q_0 is defined to be a 0 when $A = 0, B = 0, C = 0, D = 0$. Q_1 is defined to be a 0 when $A = 0$, $B = 0, C = 0, D = 1$. Q_2 is defined to be a 0 when $A = 0, B = 0, C = 1, D = 0$. Q_3 is defined to be a 0 when $A = 0, B = 0, C = 1, D = 1$.

```
module decoder
title '4-to-8-line_decoder'
decoder device 'P16L8';
@ALTERNATE

A,B,C,D             pin 1,2,3,4;
/Q0,/Q1,/Q2,/Q3     pin 12,13,14,15;

equations

/Q0 = /A * /B * /C * /D;
/Q1 = /A * /B * /C * D;
/Q2 = /A * /B * C * /D;
/Q3 = /A * /B * C * D;

end decoder
```

Figure 6.16 ABEL program for decoder

The functions are:

$$\bar{Q}_0 = \bar{A}\bar{B}\bar{C}\bar{D}$$
$$\bar{Q}_1 = \bar{A}\bar{B}\bar{C}D$$
$$\bar{Q}_2 = \bar{A}\bar{B}C\bar{D}$$
$$\bar{Q}_3 = \bar{A}\bar{B}CD$$

Clearly these functions will easily fit into a PAL such as PAL16L8. The functions are described in the ABEL program shown in Figure 6.16. The PAL16L8 device is specified in this program. Each input and output is associated with an appropriate device pin in the declaration section. Notice that because we are using a PLD with inverted (active-low) outputs, the pin names also include the inversion symbol. The inversion symbols also appear with outputs since the equations to be implemented specify output inversion.

Of course, we probably would not use a PLD for a simple decoder application as standard MSI parts exist that could be used (e.g. the 74LS154 4-line-to-16-line decoder). But suppose the decoder required an output to be a 0 when one of two combinations of input were present, say $\bar{A}\,\bar{B}\,\bar{C}D$ or $\bar{A}B\,CD$. Then we would have a sum-of-product expression for the output, e.g.:

$$\bar{Q}_0 = \bar{A}\,\bar{B}\,\bar{C}D + \bar{A}B\,CD$$

which can be written in the equation section as:

```
/Q0 = /A * /B * /C * D + /A * B * C * D;
```

ABEL offers several features to aid the specification. For example, the equations need not be in the format actually required by the PLD. A Boolean expression can be written using the symbols of ABEL in any way which is a valid Boolean expression (including the use of parentheses to enforce precedence), and the ABEL compiler will attempt to "reduce" the expression into a suitable form for the specified PLD. It may not be immediately obvious how many terms will be present in the final form, and it may be that it will not be possible to use the selected PLD.

One of the most common rearrangements is to complement the output function. For example, the function:

$$f = \bar{A}\bar{B}\bar{C}D + \bar{A}BCD$$

is specified. In ABEL, it becomes:

```
Q0 = /A * /B * /C * D + /A * B * C * D;
```

in the equation section. If we were to use a PAL16L8, the ABEL compiler would have to form the inverse expression, which has fourteen product terms when not minimized (since sixteen product terms of four variables). The final minimized function, as given by the 0's on the Karnaugh map in Figure 6.17, is:

$$\bar{Q}_0 = A + \bar{D} + B\bar{C} + \bar{B}C$$

which would fit into a PAL16L8. (A PAL with active-high outputs could be used with the original equation, but generally devices with active-low outputs are more widely available.)

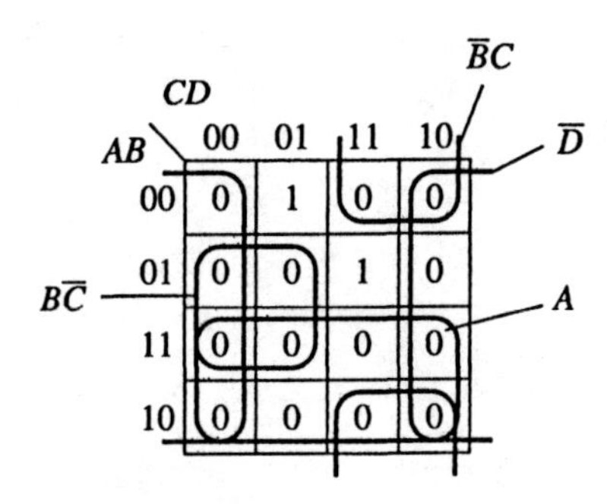

Figure 6.17 Mapping zeros to get complement function

Sets and relational operators

A set in ABEL is a group of signals given a collective name. This name can then be used within the equation section to replace the individual signal names. Sets are defined using a set declaration. For example, suppose the inputs I_0, I_1, I_2 and I_3 are to be formed into a set called X, another group of inputs, I_4, I_5, I_6 and I_7 form the set Y, and the group of outputs O_0, O_1, O_2, O_3 form the set Z. The set declarations in the declaration section of the ABEL source program are:

```
X = [I0, I1, I2, I3];
Y = [I4, I5, I6, I7];
Z = [O0, O1, O2, O3];
```

Square brackets and semicolon termination are used here. Once sets are defined, they can be used with most ABEL operators. In such cases, the sets must be of the same size and the operators apply to corresponding members of each set. For example:

```
Z = X + Y;
```

would be equivalent to:

```
O0 = I0 + I4;
O1 = I1 + I5;
O2 = I2 + I6;
O3 = I3 + I7;
```

One useful feature of the language is a group of relational operators which are used with sets. The following relational operators are available: == (equals), != (not equals), < (less than), <= (less than or equal), > (greater than) and >= (greater than or equal). Relational operators can be used between Boolean expressions, and return the expression that makes the relation true.

EXAMPLE

Suppose we want an output P to be a 1 (true) when X is less than Y (i.e. the 4-bit number applied to I_0, I_1, I_2, I_3 is less than the 4-bit number applied to I_4, I_5, I_6, I_7), we can write:

```
P = X < Y;
```

or:

```
P = [I0, I1, I2, I3] < [I4, I5, I6, I7];
```

This equation will expand into the corresponding sum-of-product expression (a rather large equation with a large number of eight-variable product terms).

SELF-ASSESSMENT QUESTION 6.4

How many product terms are there in the previous example?

Relational operators can be used in equations with normal Boolean operators. For example, to generate a signal Q to be a 1 only if the binary number on X (I_0, I_1, I_2, I_3) is 0100 or 0101, i.e. 4 or 5, we could write:

```
P = (X == 4) + (X == 5);
```

where 4 and 5 are decimal numbers. Other bases can be used, for example hexadecimal is specified by prefixing the number with ^h. Such constructions are useful for specifying memory address decode circuits which have to respond to specific numbers on the address signals.

Constants can be declared in the declaration section, and so we might write:

```
limit1 = 4; "constant declaration"

P = (X == limit1);
```

The equality symbol is used in both constant declarations and equation assignment statements.

The set notation also provides for a range without writing the intermediate signals, using the notation `[x0 .. x5]` for `[x0, x1, x2, x3, x4, x5]`. This notation is suitable if a group of signals are used with "indices".

EXAMPLE

Suppose a circuit has a sets of six inputs, A_0, A_1, A_2, A_3, A_4, A_5, and a set of six outputs Q_0, Q_1, Q_2, Q_3, Q_4, Q_5. The inputs, A_0, A_1, A_2, A_3, A_4, A_5, are passed to the outputs, Q_0, Q_1, Q_2, Q_3, Q_4, Q_5, when an input, C, is a 0. $\bar{A}_0$, $\bar{A}_1$, $\bar{A}_2$, $\bar{A}_3$, $\bar{A}_4$, $\bar{A}_5$ is passed to the outputs, Q_0, Q_1, Q_2, Q_3, Q_4, Q_5, when the C input is a 1. The ABEL program is shown in Figure 6.18. The ABEL compiler has to expand the set notation into individual equations for each output. In this example, the statement:

```
Q = /C * A + C * /A;
```

is equivalent to:

```
Q0 = /C * A0 + C * /A0;
Q1 = /C * A1 + C * /A1;
Q2 = /C * A2 + C * /A2;
Q3 = /C * A3 + C * /A3;
Q4 = /C * A4 + C * /A4;
```

```
module complementer
title 'complementer'
decoder device 'P16L8';
@ALTERNATE

A0, A1, A2, A3, A4 , A5  pin 2, 3, 4, 5, 6, 7;
Q0, Q1, Q2, Q3, Q4, Q5   pin 12,13,14, 15, 16, 17;
C                        pin 1;

equations

Q = /C * A + C * /A;

end
```

Figure 6.18 ABEL program for complementer

The set notation is particularly convenient when dealing with collections of "bus" signals found in microprocessor systems.

Truth table method

Rather than write the Boolean equations for combinational functions, ABEL provides the truth table method in which each value of the function is enumerated. The general form starts with the keyword `truth_table` and lists in parentheses the names of the inputs and outputs, thus creating the heading for the truth table. On subsequent lines, the values of the inputs and outputs are listed. For example, we might have truth table having three inputs, A, B, C and three outputs O_0, O_1 and O_2, defined as:

```
truth_table ([A, B, C] -> [O0, O1, O2])
             [0, 0, 0] -> [ 0,  0,  0];
             [0, 0, 1] -> [ 1,  0,  1];
             [0, 1, 0] -> [ 0,  0,  0];
             [0, 1, 1] -> [ 0,  1,  1];
             [1, 0, 0] -> [ 1,  1,  0];
             [1, 0, 1] -> [ 0,  1,  1];
```

This particular truth table is incomplete (ABC = 110 and 111 not given). To make these entries interpreted as don't cares, we place the `@DCSET` directive (don't care set) earlier in the program.

SELF-ASSESSMENT QUESTION 6.5

What set of Boolean functions does the previous truth table describe?

Test vectors

The test vector section is optional but, together with the simulator, is useful to establish that the design will function according to specification. The form is the same as the truth table. After a heading, in this case using the keyword `test_vectors`, the list of inputs and required outputs is enumerated.

6.4.3 *Sequential functions*

Equation method

To program a registered PAL, in addition to specifying any combinational output function, the sequential nature of the circuit must be specified. The assignment operator is extended into a ***registered assignment operator***, :=, which is used in equations to define the outputs of flip-flops. If a flip-flop output, Q_0, is to be the same as the input, I_0, after each clock transition (i.e. a simple D-type flip-flop), we would simply write:

```
Q0 := I0;
```

in the equation section.

EXAMPLE

To design a 3-bit binary counter, we know from Chapter 4, Section 4.5.2, that the input equations for D-type flip-flops are:

$$D_A = \bar{A}$$
$$D_B = A\bar{B} + \bar{A}B$$
$$D_C = \bar{A}C + \bar{B}C + AB\bar{C}$$

where D_A, D_B and D_C are the flip-flop inputs and A, B and C are the flip-flop outputs. Hence the ABEL source program could be as shown in Figure 6.19. (Of course if all we wanted was a 3-bit counter, probably using an MSI part such as a 74LS163 4-bit counter would be more cost effective, unless the remaining part of the PLD could be used for other purposes.)

State diagram method

An alternative way of specifying a sequential circuit in ABEL is to use the state diagram format. In this format, states are specified and Boolean conditions that cause a transition to other specified states. Conditional constructs similar to IF–THEN–ELSE statements, GOTO and CASE statements are used.

```
module three-bit_counter
title 'three-bit_counter'
decoder device 'P16R8';
@ALTERNATE

A,B,C          pin 2,3,4;
Q0,Q1,Q2       pin 12,13,14;
CLK            pin 1;

equations

Q0 := /A;
Q1 := A * /B + /A * B;
Q2 := /A * C + /B * C + A * B * /C;

test_vectors ([A, B, C] -> [Q0, Q1, Q2])
              [0, 0, 0] -> [0,  0,  1];
              [0, 0, 1] -> [0,  1,  0];
              [0, 1, 0] -> [0,  1,  1];
              [0, 1, 1] -> [1,  0,  0];
              [1, 0, 0] -> [1,  0,  1];
              [1, 0, 1] -> [1,  1,  0];
              [1, 1, 0] -> [1,  1,  1];
              [1, 1, 1] -> [0,  0,  0];

end
```

Figure 6.19 ABEL program for a 3-bit binary counter

EXAMPLE

Consider the state diagram derived in Chapter 5, Section 5.2.1, for a counter which has two different repeating sequences dependent upon a control input, named *P* here. The ABEL program for this counter is shown in Figure 6.20 together with the state diagram drawn on the left for reference.

Each state is given a name (rather than a number as in the previous chapters) and state variable values are assigned to states using equation statements prior to the state diagram specification. The state diagram specification starts with the keyword `state_diagram`, followed by the state variables (normally a set). Subsequent lines identify a state, the corresponding outputs and an expression specifying the transitions to the next state(s). In our example, there are no specifically generated outputs; the flip-flop outputs are the outputs of the circuit.

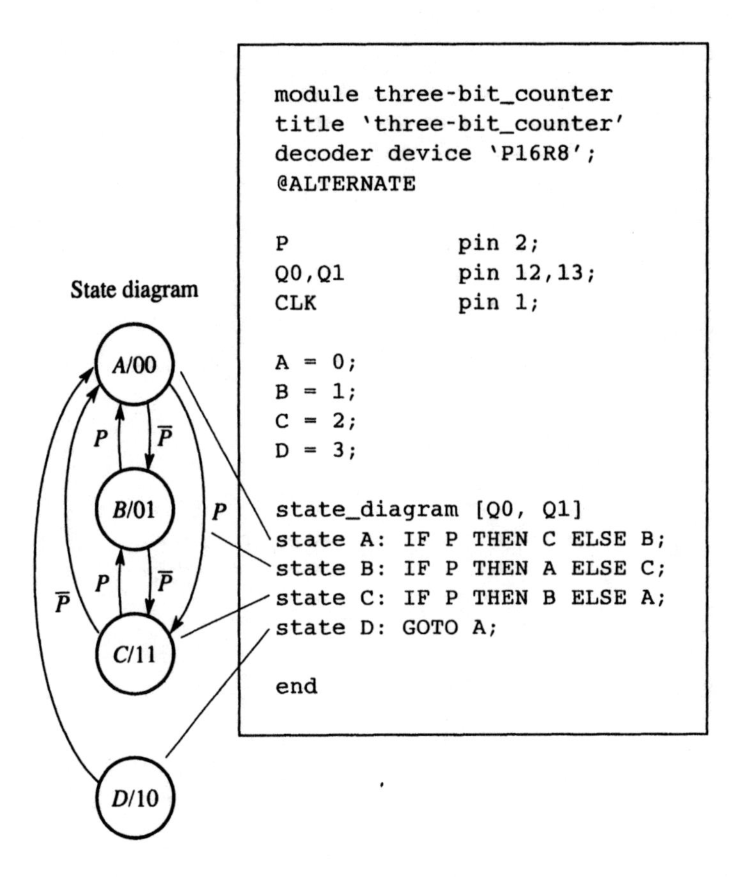

Figure 6.20 ABEL program for two-sequence counter using state diagram notation

Truth table method

A sequential circuit can be described in a truth table format which lists the current state and input values and the corresponding next state (essentially a state transition table). The truth table for the previous two-sequence counter is shown in Figure 6.21. The keyword `truth_table` heads the list, together with the names of each column, which is similar to the truth table description for combinational circuits. The only difference is the use of the symbols `:>` rather than `->`. Notice the use of the ABEL symbols `.X.` for a don't care.

Test vectors

These are listed in a similar fashion for a sequential circuit as for a combinational circuit. The reset input would be listed to force the circuit to a known state. When the activating clock transition occurs, the right hand side output should be generated.

```
module three-bit_counter
title three-bit_counter
decoder device 'P16R8';
@ALTERNATE

P               pin 2;
Q0,Q1           pin 12,13;
CLK             pin 1;
RESET           pin 3;

A = 0;
B = 1;
C = 2;
D = 3;

truth_table (P, [Q0, Q1]) :> [Q0,Q1])
             0,     A      :>   B;
             1,     A      :>   C;
             0,     B      :>   C;
             1,     B      :>   A;
             0,     C      :>   A;
             1,     C      :>   B;
            .X.,    D      :>   A;

end
```

Figure 6.21 ABEL program for two-sequence counter using truth table notation

Architecture independent designs

If we want to omit the device type and pin numbers from the previous design to delay the specific device selection, it may be necessary to resolve ambiguities. When a signal is declared as a clock input previously, it was assigned to the correct pin number of the specific device. Without this information, the clock signal has to be identified as such. This can be done with the statement:

```
[Q0,Q1].clk = clock;
```

which says that the clock input for the flip-flops Q_0 and Q_1 will be called `clock`. The extension, `.clk` is a dot extension. Similarly, it may be necessary to specify that Q_0 and Q_1 are outputs of a certain type of flip flop, say *D*-type flip-flops. The `Istype` attribute statement can be use for this purpose, i.e. :

```
[Q0, Q1] Istype 'reg_d';
```

where `'reg_d'` is an ABEL attribute specifying *D*-type outputs.

6.5 Using read-only memories (ROMs)

Let us complete this chapter with a discussion of read-only semiconductor memory which strictly is not a PLD but can be programmed as a combinational logic circuit. A read-only memory (ROM) is a type of memory whose contents are normally only accessed for reading and not accessed for altering the contents (writing). Semiconductor read-only memory is also non-volatile in that the information is not lost when the power is removed. Its primary use is in computer systems to hold information that must be present when the system is switched on. However read-only memory also finds application as a component in logic design.

The basic structure of a read-only memory (as all semiconductor memory) consists of a two-dimensional array of memory cells, each memory cell storing the value of one binary digit (a 0 or a 1), as shown in Figure 6.22. The memory in this figure is organized as "×1" which means that one bit can be accessed at a time. Each cell has unique position in the array given by a row and a column. In Figure 6.22, there are 2^r rows and 2^s columns. An r-bit number can select the row and an s-bit address can select the column. The complete "address" of the cell is given by r and

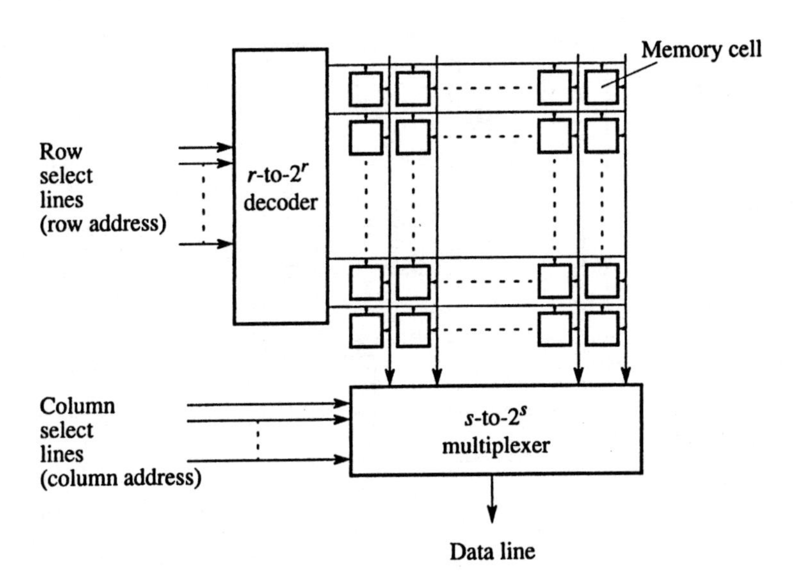

Figure 6.22 Two-dimensional structure of semiconductor memory (×1 organization)

s concatenated. For a fixed number of memory cells, the least number of rows and columns combined is obtained when the array is square.

Memory cells can be constructed in one of several ways, including the *fixed read-only memory* whose stored information can never change after manufacture. There are types of read-only memory which allow the information to be altered by erasing the stored information and writing new information into the memory cells – the so-called *programmable read-only memory*, PROM, and the *erasable programmable read-only memory*, EPROM. Writing new information is a relatively slow process which often requires the device to be removed from the system.

To access the stored information in a memory cell, the row address is provided which selects all the cells on a row. A column is selected with the column address to finally select the cell. For reading, the information passes from the cell down the column to a single output pin. (For writing, when this is possible, the information passes in the opposite direction into the selected cell.)

Often cells are organized into sets of words, and the bits of a complete word are accessed simultaneously. This can be achieved by duplicating the $\times 1$ organization for each bit of the word as shown in Figure 6.23, which is the so-called $\times n$ organization when there are n bits in each word. In this design, a single row decoder is used, but separate column select circuitry is used for each bit of the word. It may be difficult to have n large in a single package because of the pin constraints. However $\times 8$ can be contained in a single package.

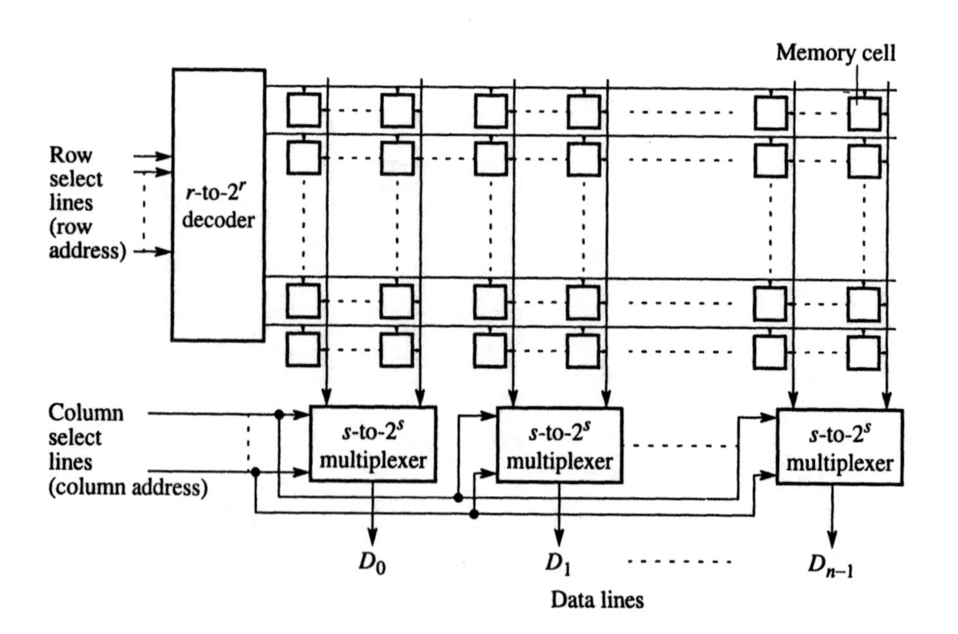

Figure 6.23 $\times n$ *organization of semiconductor memory*

SELF-ASSESSMENT QUESTION 6.6

What is the largest ×1 memory organization that can be contained in a 20-pin package?

SELF-ASSESSMENT QUESTION 6.7

What is the largest ×8 memory organization that can be contained in a 20-pin package?

For our purposes, we can consider the read-only memory as a packaged device having storage locations each with a unique address. The device has inputs (address inputs) and outputs (data outputs). Viewed as this, we can see that a set of inputs are applied and a set of outputs are generated. The output values simply depend upon the (permanent) contents of the memory cells selected by the inputs. Hence, the memory can be considered as a combinational logic circuit.

EXAMPLE

Suppose the combinational function given in Table 6.1 is to be implemented in a read-only memory. The read-only memory will have exactly the same contents of the truth table, as shown in Figure 6.24. No attempt is made to

Table 6.1 *Truth table of function for read-only implementation*

A	*B*	*C*	*D*	*f(A,B,C,D)*
0	0	0	0	0
0	0	0	1	1
0	0	1	0	0
0	0	1	1	1
0	1	0	0	0
0	1	0	1	1
0	1	1	0	1
0	1	1	1	1
1	0	0	0	0
1	0	0	1	0
1	0	1	0	1
1	0	1	1	1
1	1	0	0	1
1	1	0	1	0
1	1	1	0	0
1	1	1	1	0

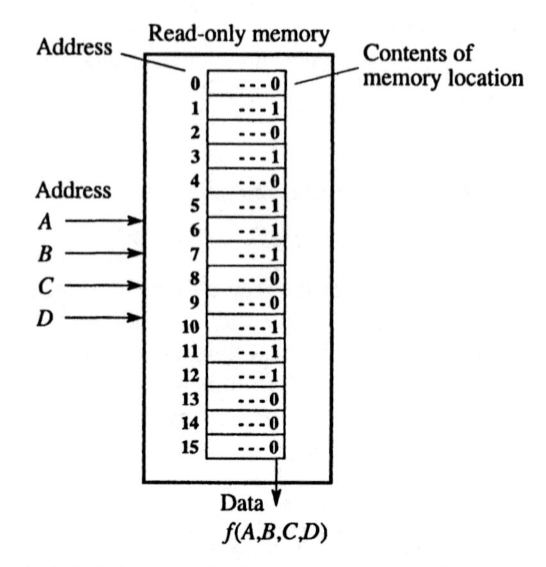

Figure 6.24 Using a read-only memory as a combinational circuit

reduce the function as this is unnecessary for this implementation. (It may be necessary if the number of variables is too great to fit into a specific read-only memory.) In Figure 6.24, only one output is used from the read-only memory as a single Boolean function is given.

As many read-only memories have a $\times n$ organization, clearly such read-only memories would be suitable for multiple functions. If $n = 8$, we could have eight independent functions of the inputs, in a similar manner as the independent functions that are possible from PLDs. Even if we do not use all n outputs, it still may be a cost-effective solution. One disadvantage of using read-only memories is that they tend to be much slower in operation than PLDs or basic gates.

6.6 Summary

This chapter has covered:

- The concept of a PLD.
- The programmable logic array (PLA).
- The popular programmable array logic (PAL).
- Sequential PLDs.
- Various designs that can be implemented with PLDs.
- PLD software tools, notably ABEL in some detail.
- Using read-only memory as a combinational logic circuit.

6.7 Tutorial questions

6.1 Implement the function $f = AC\overline{D} + BC + \overline{E}$ in a PLA. Show the internal connections.

6.2 Implement the function $f = AC\overline{D} + BC + \overline{E}$ in a PAL. Show the internal connections.

6.3 A sequence detector is designed in Chapter 5 (Section 5.2.2) having a single data input and a single output which becomes a 1 immediately after the sequence 0110011 appears on the input. The flip-flop input equations are:

$$Y_3 = \overline{x}\,\overline{y}_1 + \overline{x}y_3 + y_2y_1$$
$$Y_2 = \overline{x} + y_3y_2\overline{y}_1$$
$$Y_1 = y_3y_1 + y_3y_2y_1 + \overline{x}\,\overline{y}_2\overline{y}_1 + \overline{x}y_2y_1$$

The output equation is:

$$Z = y_3\overline{y}_2\overline{y}_1$$

Implement this sequence detector using a registered PAL. Show the internal connections of the PAL.

6.4 Write an ABEL program to implement a circuit which compares two 6-bit numbers, A and B, and generating three outputs, one which is a 1 when $A < B$, one which is a 1 when $A = B$, and one which is a 1 when $A > B$.

6.5 Write an ABEL program to implement the problem in Question 6.3.

6.6 Prove that the least number of rows and columns combined in a memory array is obtained when the array is square.

6.8 Suggested further reading

Advanced Micro Devices, *Product Line & Selector Guide*, Sunnyvale, CA, 1990.

Data I/O Corp., *ABEL™ Design Software User Manual*, Redmond, Washington, 1990.

Pellerin, D. and Holley, M., *Digital Design Using ABEL*, Prentice Hall: Englewood Cliffs, New Jersey, 1994.

Wakerly, J. F., *Digital Design Principles and Practices, 2nd edition*, Prentice Hall: Englewood Cliffs, New Jersey, 1994.

CHAPTER 7

Testing logic circuits

Aims and objectives

This final chapter is provided to give an overview of testing logic circuits. Basic logic testing methods will be outlined as applied to simple combinational circuits. Methods for more complex sequential circuits and systems will be briefly described.

7.1 The need for testing

Once a logic device or logic system is manufactured, it is necessary to test it to detect flaws which may have arisen during manufacture or subsequently during handling and use. (We assume that the design has been found to be logically correct through the use of a logic simulator.) Testing is of extreme importance and should be an early design consideration, as circuit fabrication is not perfect. Circuits can have physical defects (*faults*) occurring during manufacture that can alter their behaviour. Even handling a circuit afterwards can damage it. The goal of testing is to detect these faults so that faulty circuits can be identified. The results of testing can be fed back to the fabrication line so that the fabrication process can be improved.

Testing is a significant intellectual challenge because of the enormous number of possible faults in a circuit. The only way we know that a packaged component works is to apply test signals to its *primary inputs* (the inputs to the packaged component) and observe the *primary outputs* (the outputs of the packaged component). If the outputs are different from what is expected, we assume a fault or design error exists. Design errors are detected by applying test vectors to a simulator. Test vectors are based upon the required function. For *fault* testing, we apply test vectors to recognize particular faults in the device. How easy it is to recognize a faulty device will depend upon the degree to which internal circuit nodes can be *controlled* and *observed*. Often, we do not need to find the specific location of the fault, but only that a fault exists in the device (unless a particular feature of manufacturing is being investigated).

7.2 Faults and fault models

To test for faults, we first need to know the types of fault that can occur in a circuit. There are many types of physical fault that can exist and affect the operation of the circuit. To cope more easily with the complexity of testing, it is convenient to provide a model of faults. In general, faults can be modelled by the effect they have on the functionality of the circuit. Fault models can be divided into two broad categories, logic fault models and parametric fault models. Logic fault models deal with faults that affect the logic function of the circuit. Parametric fault models deal with faults that affect the magnitude of circuit parameters such as voltage, current, drive and delay. Here we shall only consider logic fault models.

Stuck-at fault

Many faults can be modelled as causing a permanent logic 1 or a permanent logic 0 on a signal path (input or output of a gate). A fault leading to a permanent 1 on a signal path is said to be *stuck-at-1* (s-a-1), and a fault leading to a permanent 0 is said to be *stuck-at-0* (s-a-0).

Other types of fault

Apart from a signal being permanently forced to a logic 1 or a logic 0, other situations arise. Sometimes an internal short or open may not create a permanent 0 or permanent 1 output, but rather an output at an indeterminate intermediate voltage. This type of fault is known as a *stuck-on fault*, or a transistor level stuck-at fault. Such faults are difficult to detect, and we shall not consider them here.

A type of fault that can exist in CMOS logic circuits which cannot be modelled by the stuck-at model is the *stuck-open fault*. In this fault, a transistor behaves as an open circuit (high impedance) permanently. This can cause the output to maintain its previous value for short periods of time (usually a few milliseconds) when both push-down and pull-up transistors are off.

A short-circuit between two parts of a circuit is called a *bridging fault*. The actual effect on the gate will be dependent upon the technology of the gate and the specific location. It may change the Boolean function or behave as an stuck-at fault. Other effects exist. For example, a bridging fault between the input and the output of a combinational logic circuit might cause a sequential circuit to be formed. When two outputs are bridged, the resultant voltage of the lines connected together will depend upon the relative power of the gates to drive the lines low and high.

Single stuck-at fault model

The stuck-at fault model is the most popular because of its simplicity. It is used as the main fault model in most fault simulators. We shall concentrate upon this model here. Figure 7.1 shows the sites of stuck-at faults in a two-input AND gate and the resulting output. The actual fault is probably inside the gate. The stuck-at model simply gives the effect of the physical fault on the inputs/outputs.

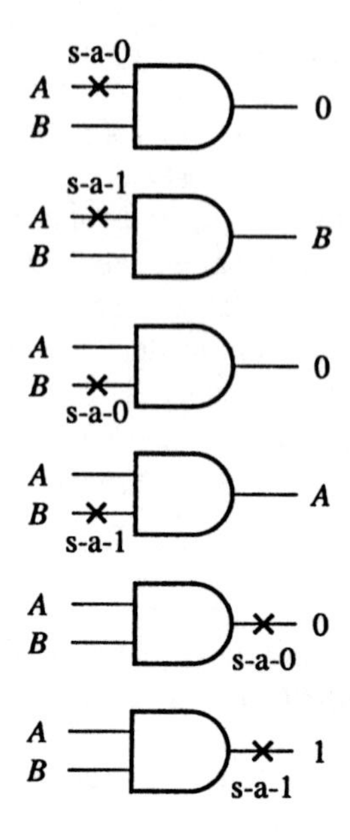

Figure 7.1 Stuck-at faults in an AND gate

SELF-ASSESSMENT QUESTION 7.1

Draw a figure similar to Figure 7.1 but for the stuck-at faults in an OR gate.

In a two-input AND gate, there are six possible single stuck-at faults, one of two faults on each of three lines. In general with k signal lines, there are $2k$ different combinations of single faults. For multiple faults, each signal line could be fault-free, stuck-at-0, or stuck-at-1 (three states). Given k distinct lines for faults, there are 3^k possible combinations of the three states on each line. One combination is the fault-free condition for the circuit. Therefore there are $3^k - 1$ different fault conditions. There could be a very large number of combinations of faults in a circuit. We shall only consider the possibility of single faults and not multiple faults because of the complexity of testing multiple faults explicitly. Fortunately a large percentage of multiple faults are actually detected during testing for single faults.

Figure 7.2 shows a simple logic circuit. This circuit has three primary inputs, A, B and C, and one primary output, f. There is one internal signal path, marked x. A fault could manifest itself on any the signal lines: In Figure 7.2, there are five possible sites for faults, A, B, C, x and f, as marked with crosses. Each site could have a stuck-at-1 fault or stuck-at-0 fault. Hence there are 10 possible single stuck-at faults in this circuit.

If there is connection from a gate output to more than one gate input, each path needs to be considered separately, as the stuck-at fault model represents the manifes-

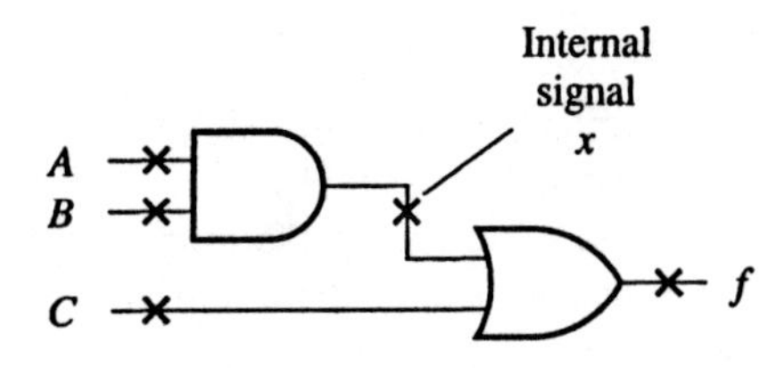

Figure 7.2 Location of stuck-at faults in a simple logic circuit

tation of some internal fault which might not be a direct fault on the signal line affecting all the inputs connected to the output. Figure 7.3 shows a circuit with an output connected to two inputs. In this case, we have eight different stuck-at sites. Now the output of the AND gate could have an output stuck-at fault which affects both the input of the OR gate and NOT gate. The OR gate and NOT gate could have input stuck-at fault which only affects themselves.

7.3 Generating test vectors for combinational circuits

A very simple circuit could be tested by applying all input values and comparing the output to the truth table of the circuit. For example, a two-input AND gate could be fully tested by applying 00, 01, 10 and 11 to the inputs and observing that the output is 0, 0, 0 and 1 respectively. If the output is different in any instance, we know for certain that there is a fault. The actual input patterns used for detecting faults are known as the *test vectors*. The group of test vectors used to test the circuit is called the *test set*. For example, the exhaustive approach on a two-input AND gate uses the test vectors 00, 01, 10 and 11 (the test set). The number of combinations of input values, of course, is directly related to the number of inputs. There are 2^n different input combinations with n inputs. Hence, as the complexity of the circuit and the number of inputs increase, it becomes impractical to apply all combinations of input values.

The *fault coverage* of a test is the fraction of faults that can be detected by the test. The central problem is how to choose the smallest number of test vectors for a given fault coverage.

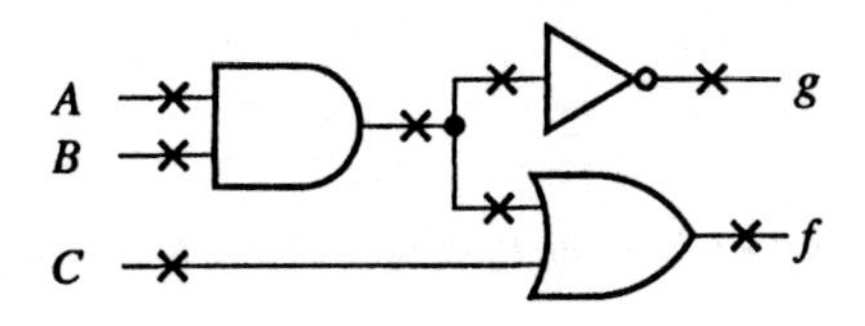

Figure 7.3 Location of stuck-at faults with circuit having fan-out

7.3.1 Tabular and algebraic methods

If we know the Boolean function of the circuit, a simple comparison of the function without a fault and with a fault can lead to the Boolean expression of the test vectors. This can be done using a truth table listing the output with and without a fault. Certain algebraic manipulations can also be done starting with the Boolean function which can lead to a test set (for example, in the Boolean difference method). We shall not consider these methods because they are only suitable for small circuits that can be described by a simple Boolean function. Interested readers should consult Lala (1996).

7.3.2 Path sensitization method

The *path sensitization method* is an appropriate method for more complex combinational logic circuits. In this method, we find whether a fault exists at one point by applying the opposite value to that generated by the fault at that point. If the test is for a s-a-1 fault, we apply a 0. If the test is for a s-a-0 fault, we apply a 1. By applying suitable input values, a "sensitized" path is created through the network to propagate the fault level from the site of the fault to the output where it can be observed.

EXAMPLE

Figure 7.4 shows an example of the technique. Here the fault to detect is a s-a-0 fault at the output of an AND gate. We must attempt to drive the fault location to a 1 (the opposite to the fault level). This can be done by applying a 1 on the input of the AND gate. Then we have to propagate the signal to the primary output, which can be done by applying a 0 on the input of the OR gate. The logic level on the output of the AND gate (a logic 0 if a fault or logic 1 if without a fault) will now appear on the output of the OR gate.

On some circuits, the logic level may be inverted – for example, with the use of a NOR gate instead of a OR gate, as shown in Figure 7.4. To facilitate describing the actions, the letter D is used to indicate that the output without a fault is a 1 and with a fault is a 0. Similarly $\overline{D}$ indicates that the output is a 0 without a fault and a 1 with a fault.[1] In Figure 7.4, the output f will be D.

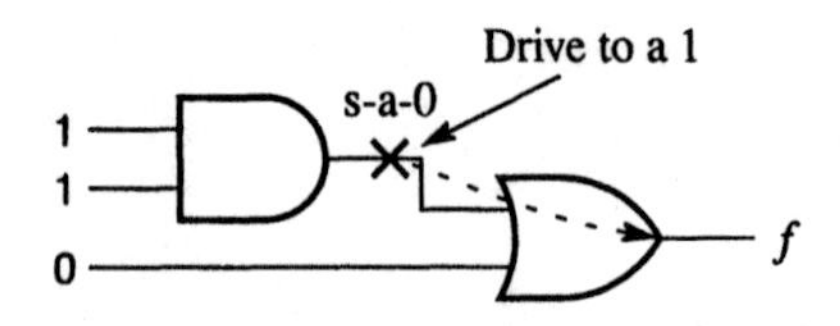

Figure 7.4 Example of using path sensitization for testing for a fault

[1] The symbols D and $\overline{D}$ with the same definitions were originally used in the D-algorithm (Roth *et al.*,

Setting the output of basic gates to a logic value

Figure 7.5 shows the signals required to set the output of each of the basic gates to a 0 or a 1. X is a don't care; either a 0 or a 1 will work. The basic gates are symmetrical in that inputs can be interchanged without changing the function.

Figure 7.6 shows the required signals to propagate a fault on the input of a gate through basic gates. The exclusive-OR and exclusive-NOR gates have the useful characteristic that the fault is propagated through to the output irrespective of the value on the other input. (Of course, we need to know if the value to look for is D or $\overline{D}$.)

Sensitizing a path

Each gate in the path to the primary output needs to be *sensitized*. The same form of input conditioning is required as in Figure 7.6, only now we propagate D or $\overline{D}$ as shown in Figure 7.7. Notice that path sensitization in Figure 7.6 and Figure 7.7 is achieved by applying 1's to the inputs of AND and NAND gates, and 0's to the inputs of OR and NOR gates (not including the site of a fault). Any values can be used for the inputs of exclusive-OR/NOR gates.

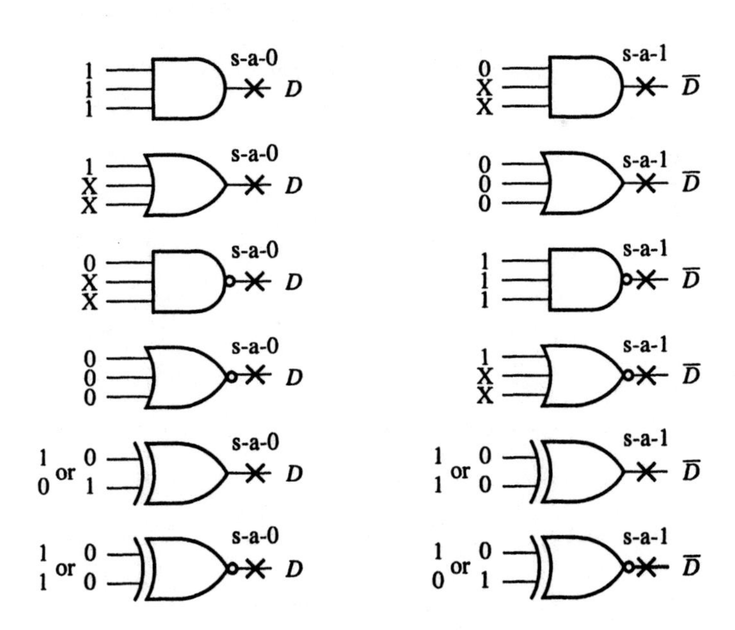

Figure 7.5 Setting output of basic gates to a logic 0 or logic 1

1967), an algorithmic way of obtaining the test vectors through path sensitization. Using the letter D does mean that it cannot be used as a normal signal name.

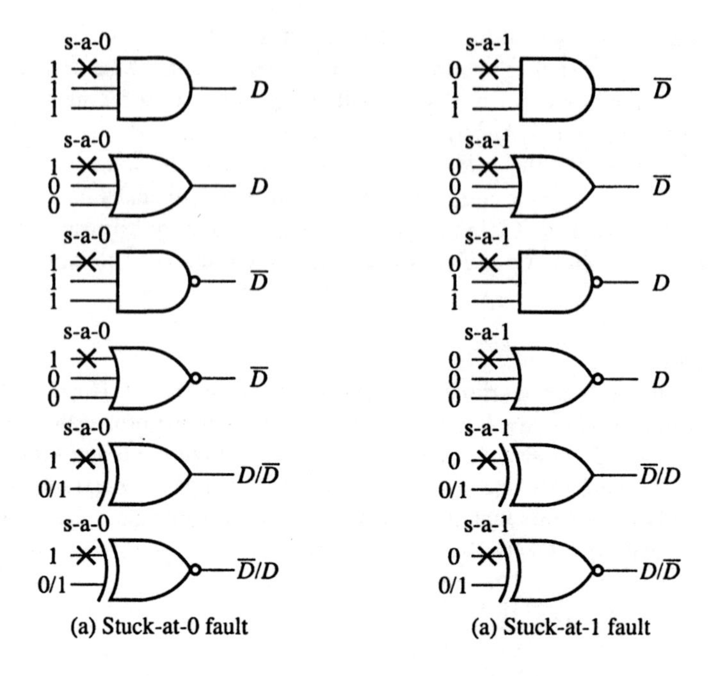

Figure 7.6 Path sensitization through gates

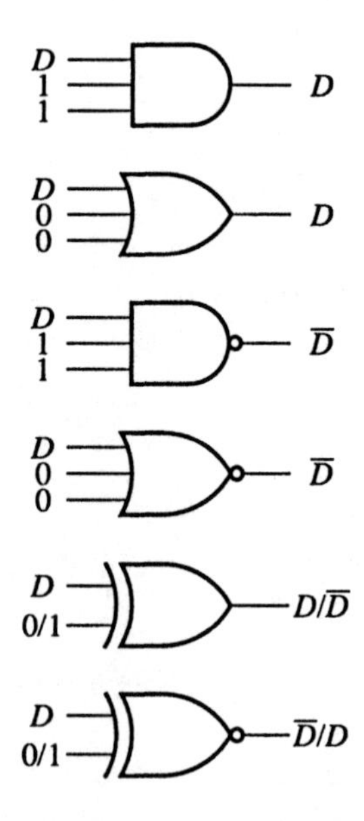

Figure 7.7 Path sensitization of $D/\overline{D}$ through gates

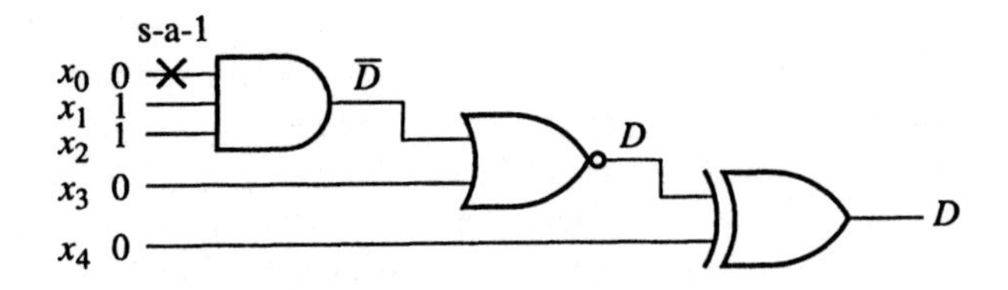

Figure 7.8 Logic circuit with path sensitization

EXAMPLE

Applying the path sensitization technique to the circuit shown in Figure 7.8 to detect a s-a-1 fault at a primary input as shown, we need to apply the values $x_0 = 0, x_1 = 1, x_2 = 1, x_3 = 0, x_4 = 0$, to obtain an observable fault at the primary output. The signal $x_0 = 0$ attempts to force the site of the fault to a 0, and the other primary signals set up a sensitized path to the primary output. If the output is a 1, we know the fault does not exist and if the output is a 0, we know a fault does exist.

Sensitizing a path – forward and backward traces

In Figure 7.8, the primary inputs could sensitize all the gates in the path to the primary output. Normally this is not the case, and we have to establish logic values on the inputs of internal gates. This can only be done by applying the appropriate values on the primary inputs. The general procedure is to first determine the logic values necessary on the inputs of each gate in the path as we propagate $D/\overline{D}$ through the circuit. In a second phase, we then go back along the path specifying the logic values of each gate before those in the path. Hence we have a two-step process, a *forward trace* and a *backward trace* in establishing the sensitized path.

EXAMPLE

For example, consider the circuit shown in Figure 7.9. The s-a-0 fault is to be detected at the point shown within the circuit. First we attempt to drive the fault

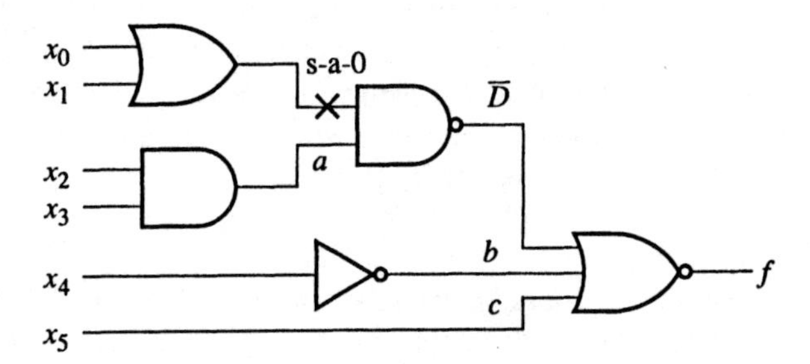

Figure 7.9 Detecting a stuck-at fault requiring a forward trace and a backward trace

site to the opposite logic level to the fault; in our case a logic 1. To do, this we must have $x_0 = 1$ or $x_1 = 1$, or both $x_0 = 1$ and $x_1 = 1$. In the forward trace, we have to sensitize a path to the output, by setting the internal point, $a = 1$, which creates $\overline{D}$ at the output of the NAND gate. To propagate $\overline{D}$ to the output, f, we need $b = 0$ and $c = 0$. In the backward trace, we work back towards the inputs. Points b and c can be set to the required value, 0, by setting the input $x_4 = 1$ and x_5 to 0. Point a can be set to 1 by setting the inputs x_2 and x_3 both to a 1. Hence the input test vector is $x_0x_1x_2x_3x_4x_5 = 011110$ or 101110 or 111110.

SELF-ASSESSMENT QUESTION 7.2

What output value in Figure 7.9 indicates a s-a-0 fault at the site shown?

Equivalent faults in a gate

Equivalent faults are those faults detected by the same test vectors. They are not equivalent if there is at least one test vector that will detect one fault but not the others. For the sake of convenience, let us consider gates with a maximum of three inputs. (The results extend naturally to any number of inputs.) Suppose a three-input AND gate is to be tested for a s-a-0 fault on the input. The test is 111 from Figure 7.6, i.e. it does not matter which input is tested for the fault. Hence all s-a-0 input faults on an AND gate are equivalent. The test vector 111 also tests for the output s-a-0 (see Figure 7.5). This leads to all s-a-0 faults on input or output of an AND gate being equivalent and we only have to test for one of them. In fact, our test vector will not be able to distinguish between the four faults. Similarly, in a three-input OR gate the test vector 000 will test for a s-a-1 fault on any input and will test for a s-a-1 fault on the output and so all input s-a-1 faults and the output s-a-1 fault are equivalent. Similarly, equivalent faults can be found in NAND and NOR gates.

SELF-ASSESSMENT QUESTION 7.3

Determine the equivalent faults for NAND and NOR gates.

There are no equivalent faults in exclusive-OR/NOR gates from Figure 7.5 and Figure 7.6. For faults to be equivalent, the complete test set for each fault must be the same. Even though the test 11 will detect both s-a-0 faults on each of the inputs of an exclusive-OR gate, they are not equivalent because another test (01) will detect one s-a-0 fault and another test (10) will detect the other s-a-0 fault. Similarly, an input s-a-0 fault and an output s-a-1 fault are not equivalent because another test vector will detect the output s-a-1 fault, namely 00, which will not detect the input faults. Similar arguments follow for the exclusive-NOR gate.

Fully testing a basic gate

To test a three-input AND gate fully, we only need to apply the test vectors 111 (to test for all input s-a-0 faults and the output s-a-0 fault) and the test vectors 011, 101, 110 (to test for input s-a-1 faults and, in consequence, the output s-a-1 fault). To test a three-input OR gate fully, we only need to apply the test vectors 000 (to test for all input s-a-1 faults and, in consequence, the output s-a-1 fault) and the test vectors 100, 010, 001 (to test for input s-a-0 faults and, in consequence, the output s-a-0 fault). In general an n-input gate can be tested with $n + 1$ test vectors even though there are $2n + 2$ different faults. Notice that if all input faults are tested, then all output faults are also tested. This applies to all basic gates. Hence, only faults on the inputs need be considered.

SELF-ASSESSMENT QUESTION 7.4

Determine the test set to test a three-input NAND gate fully and to test a three-input NOR gate fully. (You will need to do Self-assessment question 7.2 first.)

Circuits without fan-out

Gate *fan-out* is the number of gate inputs that is allowed to be attached to an output of a gate. A fan-out of eight means that eight inputs can be attached to a particular output. The word *fan-out* on its own simply suggests that there is more than one input attached to an output.

If the circuit does not have *fan-out*, i.e. a signal does not pass to more than one gate input, then a fault on an input will have only one path to a primary output. By sensitizing that path, we will detect all faults along this path, as shown in Figure 7.10. All paths in the circuit to an output can be traced back to an input. Hence it will only be necessary to apply a test vector which detects faults at the primary inputs to detect all faults in the whole circuit. This fact will very significantly reduce the number of test vectors, but only applies to circuits without fan-out. Unfortunately virtually all practical logic circuits have fan-out.

SELF-ASSESSMENT QUESTION 7.5

Why in previous chapters did we promote the use of fan-out?

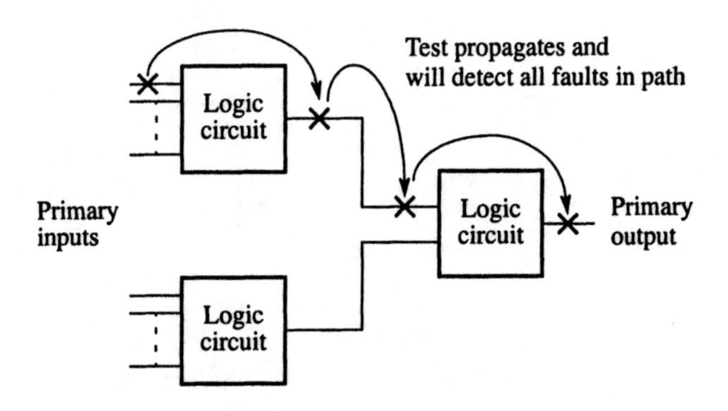

Figure 7.10 Testing input faults tests all faults

Circuits with fan-out

For circuits having fan-out, it may not be possible to test all faults in the circuit by simply testing input faults, and it may not even be possible to test input faults. While sensitizing one path, another path might become sensitized and the fault signal may pass through each path, and combine before a primary output. This combining might cause the fault signal to be indistinguishable from the fault-free signal. Figure 7.11

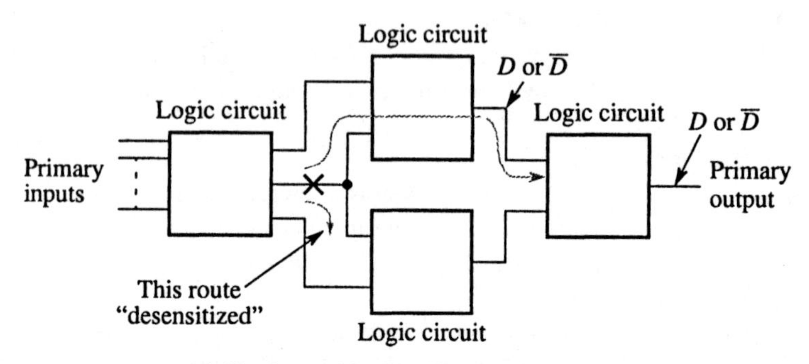

(a) Single sensitized path to primary outputs

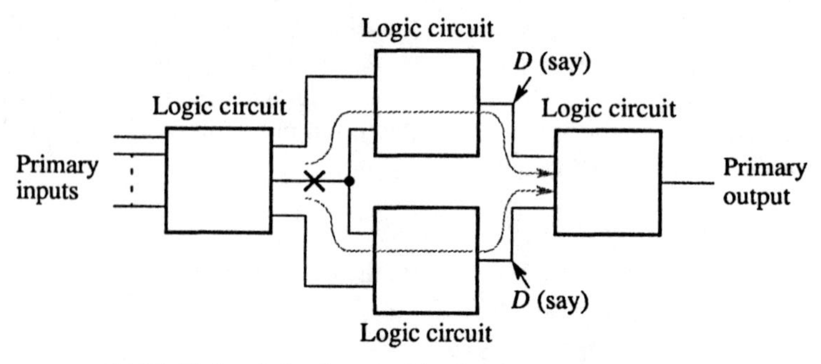

(b) Multiple reinforcing sensitized paths to primary outputs

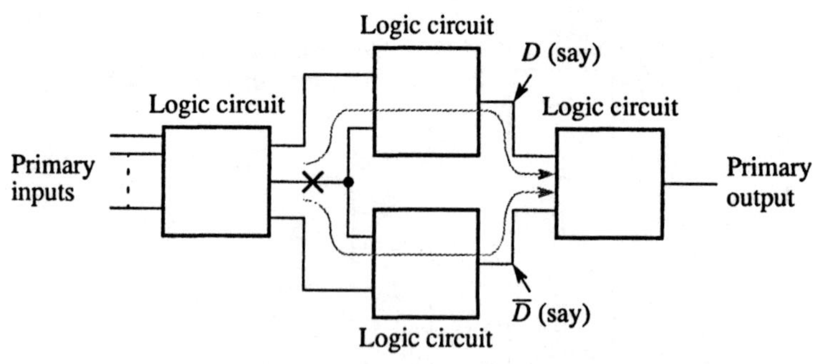

(c) Multiple cancelling sensitized paths to primary outputs

Figure 7.11 Single and multiple sensitized paths

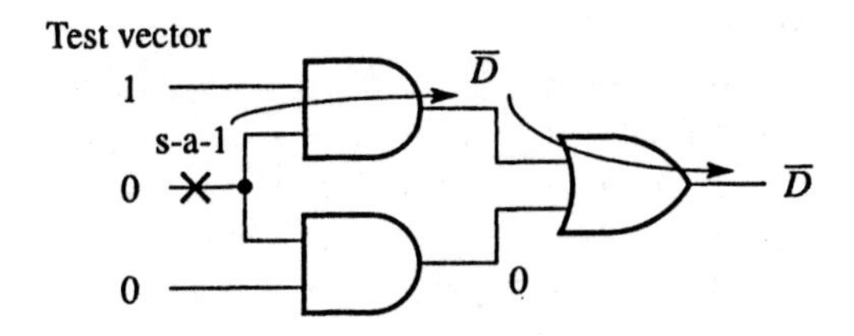

Figure 7.12 Circuit with fan-out where a single path sensitization will be satisfactory

shows the general situations that might arise. In Figure 7.11(a), there are two paths to the primary output, one of which is sensitized and one which is "desensitized". However, it may not be possible to "desensitize" one path since both may use the same signals for sensitization. In Figure 7.11(b), both paths are sensitized but the two signals interact without affecting the outcome on the primary output. For example, if two D signals are produced into an OR gate, a D single will still be produced. The fault can still be detected. However, in Figure 7.11(c), the two signals interact to affect the outcome on the primary output in such a way as to produce a signal which is not different for either with a fault or without a fault.

An example of a single path sensitization working with fan-out is shown in Figure 7.12. Here the test vector 100 will test the fault s-a-1. However, in the similar circuit shown in Figure 7.13, only a dual path sensitization will detect the fault, with the test vector 010. If we had tried a single path with the test vector 011, the path would have been blocked by a 1 on the input of the second level OR gate. In Figure 7.14, dual path sensitization causes the fault signal to cancel.

SELF-ASSESSMENT QUESTION 7.6

What would be a suitable test vector for Figure 7.14?

In general, a somewhat trial-and-error approach has to be taken to find a suitable test vector, beginning with single path sensitization. Computer programs are available for test vector selection using the path sensitization approach.

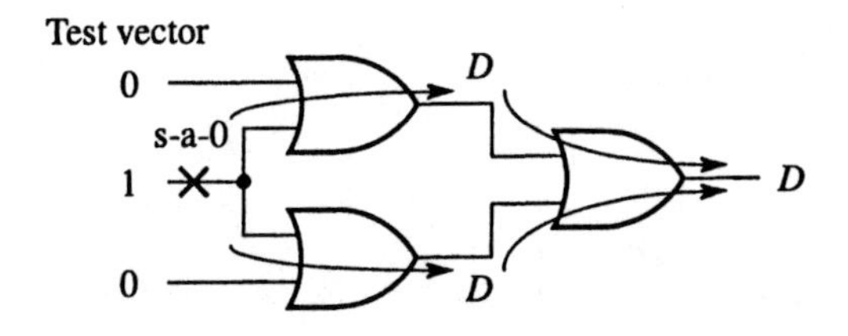

Figure 7.13 Circuit with fan-out where dual path sensitization is necessary

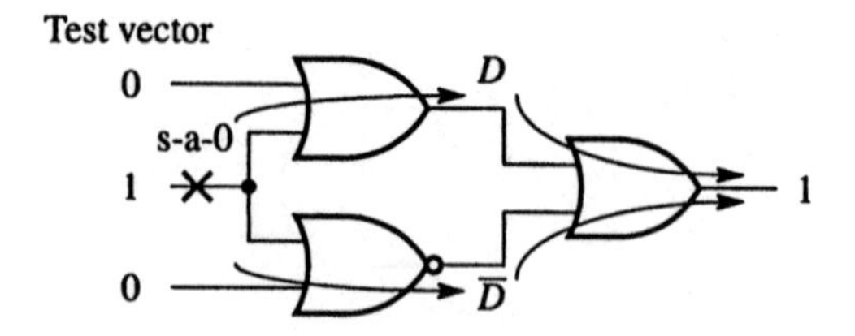

Figure 7.14 Circuit with fan-out where dual path sensitization will cause fault signals to cancel

Undetectable faults

In the previous circuit, it was possible to sensitize the path completely. In some particular circuits, the technique does not necessarily lead to a test vector. For example, consider the circuit shown in Figure 7.15. To detect the s-a-0 fault as shown, we must attempt to set the s-a-0 fault location to a 1 by setting $x_0 = 1$ and $x_2 = 1$. The path to the output is sensitized by making a and $b = 0$. To make $a = 0$, $x_0 = 0$ or $x_1 = 0$. To make $b = 0$, $x_1 = 1$ or $x_2 = 0$. These conditions conflict as x_0 and x_2 must be 1 to set the fault location to a 1. If we select $x_1 = 0$ to set a to a 0, we cannot set b to a 0. If we select $x_1 = 1$ to set b to a 0, we cannot set a to a 0. Hence the fault is "undetectable" by this method.

Other faults are also undetectable. For example, a s-a-1 fault at site b is undetectable; it requires $x_0 = 0$, $x_1 = 0$ and $x_2 = 0$ to ensure 0's on the inputs of the OR gate. To force the stuck-at fault position to a 0 requires $x_1 = 1$ and $x_2 = 0$, again a conflict.

The reason that the faults are undetectable is because this particular circuit has redundancy. In fact, the only reason that such circuits become untestable with single faults is because of redundancy in the circuit (Nelson *et al.*, 1995). By observation, the circuit implements the function:

$$f = x_0x_1 + x_0x_2 + \bar{x}_1x_2$$

But this function can be simplified to:

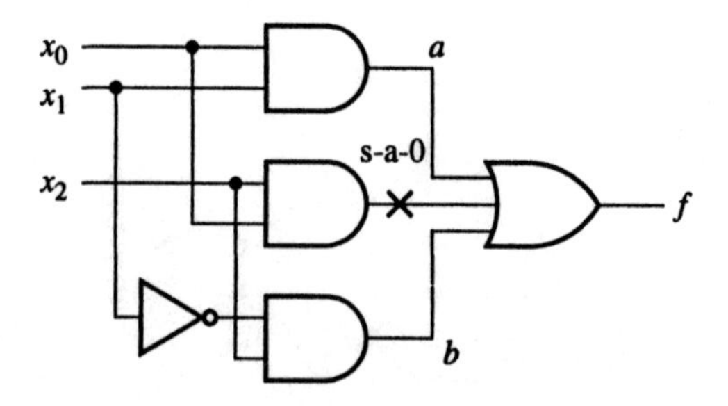

Figure 7.15 Circuit with untestable fault

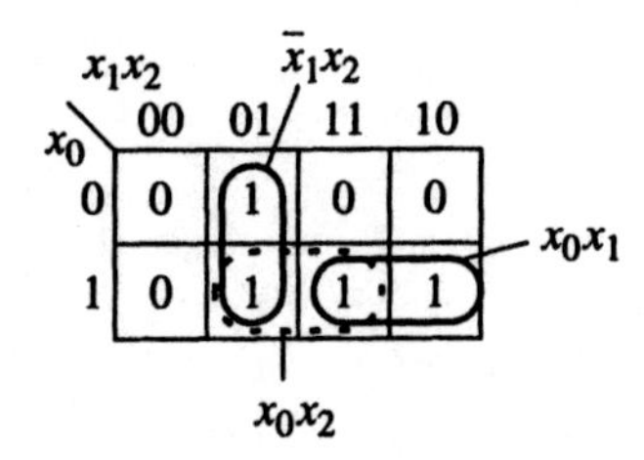

Figure 7.16 Karnaugh map of function $f = x_0x_1 + x_0x_2 + \bar{x}_1x_2$

$$f = x_0x_1 + \bar{x}_1x_2$$

from the Karnaugh map shown in Figure 7.16. We can see that the term x_0x_2 is in fact not necessary as each 1 it covers is covered by the other terms. If we remove the corresponding AND gate in the circuit, the circuit will be fully testable (the same as Figure 7.12). Though functionally unnecessary, the extra term in the original function f may have been introduced intentionally to avoid race hazards.[2] Sometimes a non-minimal expression is realized because it requires certain gates that are available, say two-input gates. If non-minimal circuits are necessary, extra outputs may need to be provided to observe internal points directly.

7.4 Testing sequential circuits and complex systems

So far we have considered testing simple combinational logic circuits. Sequential logic circuits and complete systems pose a major challenge for testing. In this section, we shall briefly review the area. Testing sequential circuits essentially consists of two parts. First, the sequential circuit must be placed in a known state. Second, the state of the circuit must be changed to another known state upon the application of input signals. This requires *controllability* of the circuit. Both the present state and the next state have to be *observable.* We presented a general model of a sequential circuit in Chapter 5 (Figure 5.1). Unfortunately the state variables may not be immediately observable. To make the state variables observable, we can use the *scan-path method.*

7.4.1 Scan-path method

In the scan-path method, we add circuits to the circuit to enable the state flip-flops to be loaded and unloaded in a serial fashion. Selectable paths are created between adjacent flip-flops to form a serial-in serial-out shift register. Serial input and output paths are made to the shift register to load and unload it, resulting in the design as shown in Figure 7.17.

---

[2] See Suggested further reading.

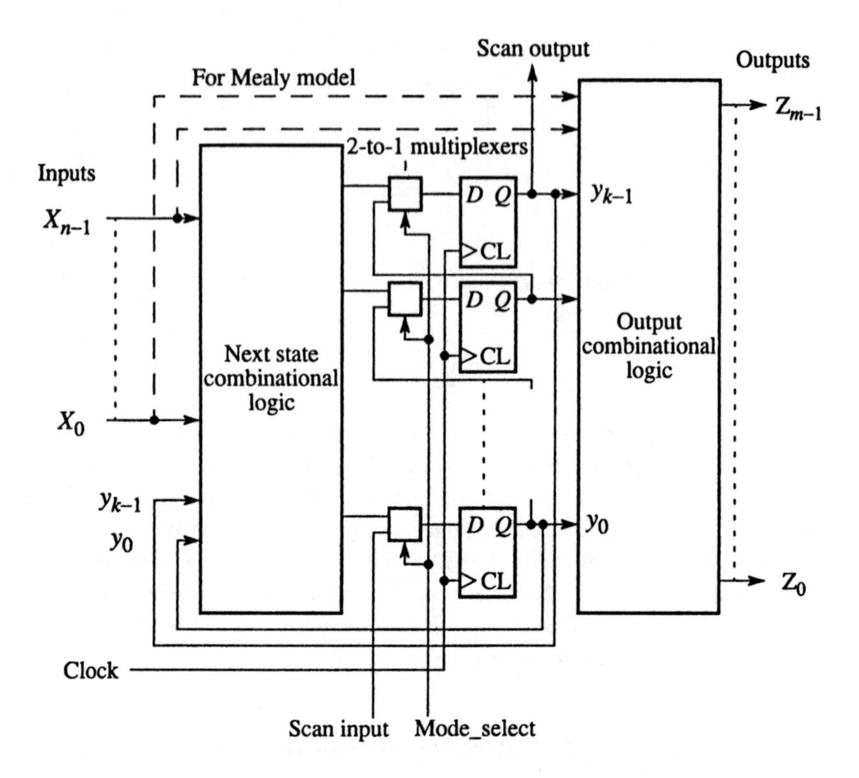

Figure 7.17 Sequential circuit model with scan paths

There are two modes of operation:

1. The normal operation in which the circuit is configured as the desired state machine. The next state logic feeds directly into the state flip-flops and the required Z outputs are generated.

2. A scan-path mode in which the state flip-flops are configured as a serial-in serial-out shift register.

The two modes can be achieved by using 2-line-to-1-line multiplexers on the inputs of the state flip-flops, as shown in Figure 7.17. A *mode_select* signal controls all the multiplexers, say *mode_select* = 0 for the normal mode and *mode_select* = 1 for scan-path mode. Each mode will be used alternately during the test.

There are three major tests that have to be performed:

1. Test the shift register configuration, i.e. test the additional logic provided for the scan-path test. This can be done by selecting the scan-path mode, shifting

into the shift register a known pattern, say 0101010 ..., and comparing this pattern with the shift register output.

2. Test the next state combinational logic. To test the next state combinational logic, we need access to both the inputs and the outputs of the logic. The inputs to the combinational logic may be available as primary inputs. The outputs of the combinational logic may be available partly as primary outputs of the circuit (if a Mealy model) and partly through the shift register when the normal mode is selected. To access these outputs, we first select the normal mode to load the flip-flops from the combinational logic and then we select the scan-path mode to shift the contents of the flip-flops out of the circuit.

3. Test the state table is properly implemented. To perform this test, we have to establish a test sequence. One approach is to drive the circuit from each of the present states as specified in the state table to the next state, i.e. a comprehensive test. This requires the state flip-flops to be loaded with the required present state values using the scan-path mode, before a specific input change to get to the next state from that present state, and then unloaded.

7.4.2 ***Built-in self-test***

In the built-in self test approach, additional circuits are provided inside the circuit not only to arrange for a test mode, but also to provide test sequences and comparators to compare with the correct output sequences. For example, the previous scan-path design could be extended into a built-in self-test by providing a sequence generator within the chip feeding into the scan input, and comparators attached to the scan output line. A pseudo-random sequence generator might be used in place of a specific sequence generator. More details of this approach can be found in Nelson *et al.*, (1995) and Lala (1996).

7.4.3 ***Boundary scan***

The boundary scan approach extends the scan path to systems of chips. Each chip is provided with a boundary scan input and a boundary scan output. Flip-flops capture information on the normal input and output pins and form a shift register, as shown in Figure 7.18. The boundary scan output of one chip connects to the boundary scan input of the next chip to form one long shift register. Chips can be individually disabled from the test by having a selectable path their shift register. This method has been adopted as an IEEE standard board test method, standard 1149.1. Not only is a shift register provided in each chip, but also additional logic for controlling the test process. Further information can be found in IEEE Standard 1149.1 (1990).

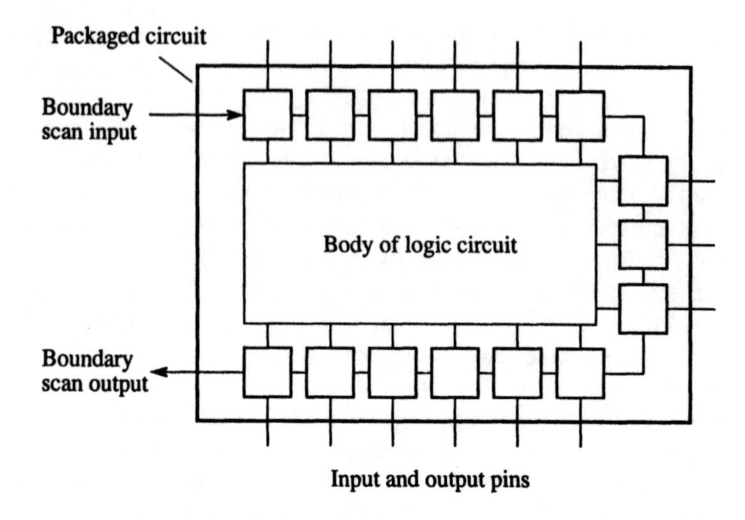

Figure 7.18 Boundary scan shift register

7.5 Summary

The following have been reviewed in this chapter:

- The concept of a stuck-at fault model
- The concept of test vectors
- The concept of path sensitization
- Testing sequential circuits (briefly).

7.6 Tutorial questions

7.1 How many stuck-at faults are there in the circuit shown in Figure 7.19?

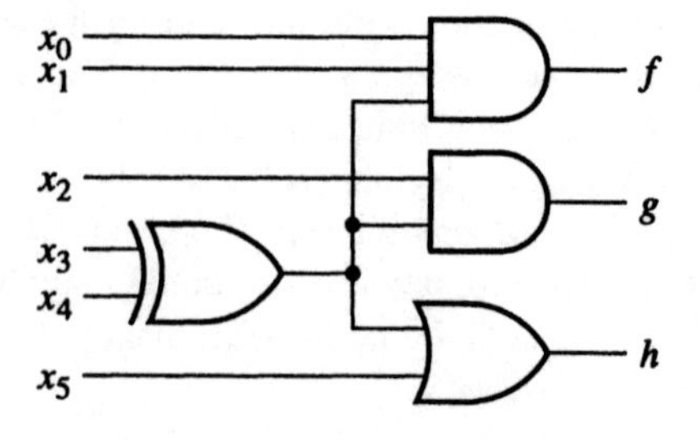

Figure 7.19 Circuit for Question 7.1

7.2 Determine the test vector which will detect the stuck-at fault shown in Figure 7.20.

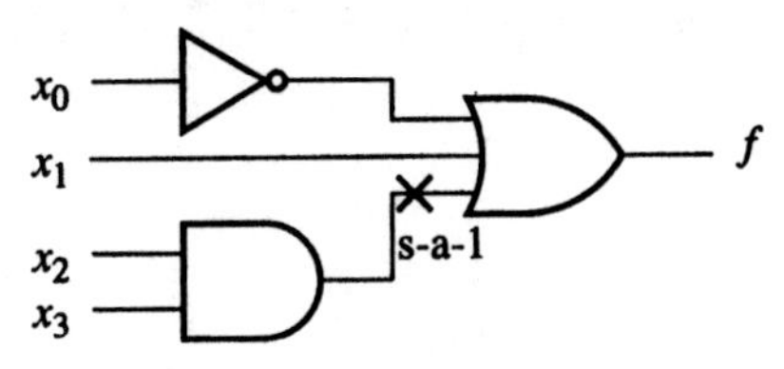

Figure 7.20 Circuit for Question 7.2

7.3 Develop the test vectors to test the logic circuits shown in Figure 7.21 for all stuck-at faults.

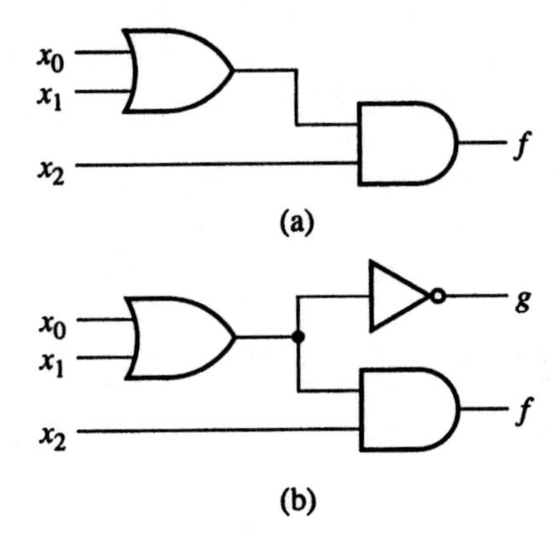

Figure 7.21 Circuits for Question 7.3

7.7 Suggested further reading

IEEE, *Test Access Port and Boundary-Scan Architecture*, IEEE Standard 1149.1, IEEE: New York, 1990.

Lala, P. K., *Practical Digital Logic Design and Testing*, Prentice Hall: Englewood Cliffs, New Jersey, 1996.

Nelson, V. P., Nagle, H. T., Carroll, B. D., and Irwin, J. D., *Digital Logic Circuit Analysis and Design*, Prentice Hall: Englewood Cliffs, New Jersey, 1995.

Roth, J. P., Bouricius, W. G. and Schneider, P. R., "Programmed algorithms to compute tests to detect and distinguish between failures in logic circuits," *IEEE Trans. Comput.*, **EC-16**, no. 5, pp. 567–80, 1967.

Wilkins, B. R., *Testing Digital Circuits An Introduction*, Van Nostrand Reinhold: Wokingham, England, 1986.

Answers to self-assessment questions

Chapter 1

1.1 $1101_2 = 13_{10}$ (decimal).

1.2 The fractional part, $0.101_2 = 0.6125_{10}$ ($0.1_2 = 0.5_{10}$, $0.001_2 = 0.125_{10}$). Hence the binary number $1101.101_2 = 13.6125_{10}$.

1.3 A pencil mark filling in a square on a questionnaire, mark = 1, no mark = 0.

1.4 255 (i.e. 11111111_2 or $2^8 - 1$)

1.5 11101010_2 (i.e. 128 + 64 + 32 + 8 + 2)

1.6 00111101_2

1.7 $42A2_{16}$

1.8

Decimal	Binary	
81	1 0 1 0 0 0 1	
63	$0_1 1_1 1_1 1_1 1_1 1_1 1$	
144	1 0 0 1 0 0 0 0	= 128 + 16

1.9 1010101 + 1 = 1010110

1.10 0101001 + 1 = 0101010

1.11 1010110. Converting back to positive using "invert digits plus 1": 0101001

1.12

Decimal	Binary
14	0 0 0 1 1 1 0
–63	1 0 0 0 0 0 1
–49	1 0 0 1 1 1 1

1.13 1110111

1.14 By changing bit b_6 from a 1 to a 0. Alternatively subtract the hexadecimal number 20_{16} from the code.

Chapter 2

2.1 A

2.2 A

2.3 A

2.4

Table 1 *Truth table to prove that A·B·1 = A·B*

A	*B*	*1*	*A·B·1*	*A·B*
0	0	1	0	0
0	1	1	0	0
1	0	1	0	0
1	1	1	1	1

2.5

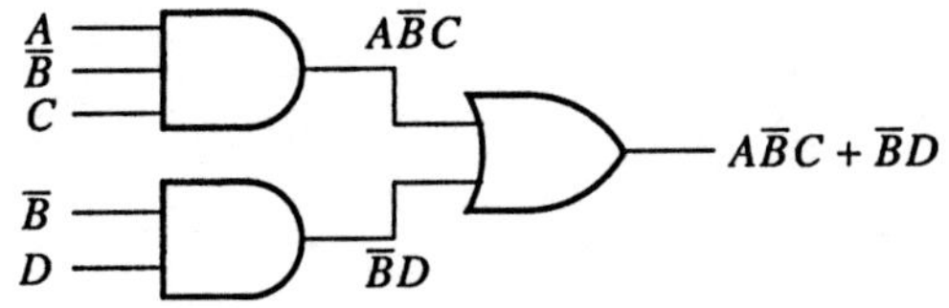

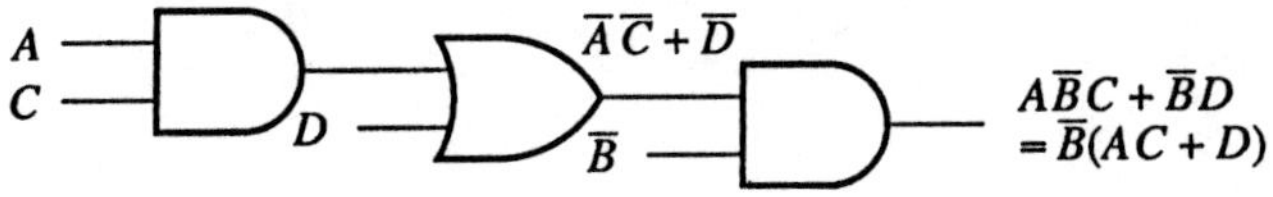

A, B, C → $A\overline{B}C$; $\overline{B}$, D → $\overline{B}D$; output $A\overline{B}C + \overline{B}D$

A, C; D → $\overline{\overline{A}\,\overline{C} + \overline{D}}$; $\overline{B}$ → $A\overline{B}C + \overline{B}D = \overline{B}(AC + D)$

2.6 $A(B + CAD) = AB + ACD$

2.7 $(A + B)(\overline{A} + C)(B + C) = (A + B)(A + C)$

2.8 First prove:

$\overline{A}\,\overline{B} = \overline{A + B}$

using a truth table, as shown in Table 2.

Table 2 *Proving* $\overline{A}\,\overline{B} = \overline{A+B}$

A	B	$\overline{A}\,\overline{B}$	$\overline{A+B}$
0	0	1	1
0	1	0	0
1	0	0	0
1	1	0	0

Then substitute $B + C$ for B in:

$$\overline{A}\,\overline{B} = \overline{A+B}$$

to obtain:

$$\overline{A}\,(\overline{B+C}) = \overline{A+(B+C)}$$
$$\overline{A}\,\overline{B}\,\overline{C} = \overline{A+B+C}$$

2.9 $f = A\overline{B}(C+\overline{D})$

2.10

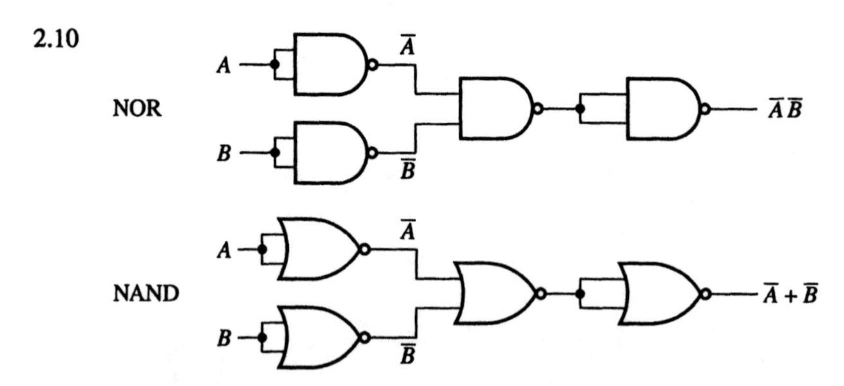

2.11 $A\overline{B}\,\overline{C} + \overline{A}B\overline{C} + \overline{A}\,\overline{B}C$

2.12 In each case, the value when $A = 0$ and $B = 1$ is different to the value when $A =$ 1 and $B = 0$.

2.13 Both upper and lower structures will be turned on causing the +5 V supply to be connected to 0 V and probably excessive current to flow through both structures. The output voltage will depend upon the internal resistances of the structures.

Chapter 3

3.1

Table 3 *Truth table of function* $f_2 = A\overline{B} + C$

A	B	C	$A\overline{B} + C$
0	0	0	0
0	0	1	1
0	1	0	0
0	1	1	1
1	0	0	1
1	0	1	1
1	1	0	0
1	1	1	1

3.2

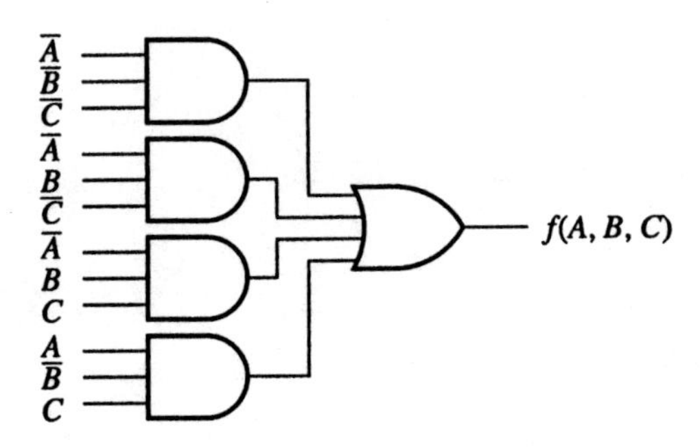

3.3 $f = \overline{A}\,\overline{B}\,\overline{C} + \overline{A}BC + ABC \quad (= \overline{A}\,\overline{B}\,\overline{C} + BC)$

3.4 First multiply A by $(B + \overline{B})$ and then by $(C + \overline{C})$.

3.5

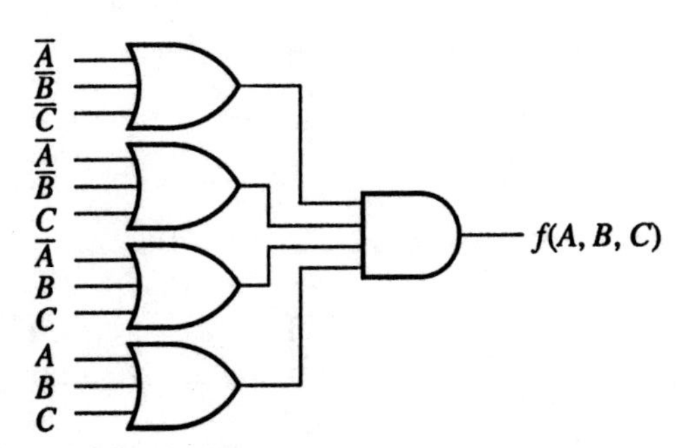

3.6 $f(A, B, C) = (\overline{A} + \overline{B} + \overline{C})(\overline{A} + B + C)(A + B + C)$

3.7 $A(BC + BD + BE + F)$

3.8 $ABC + \bar{A}DE$

3.9 $f(A, B, C) = = A\bar{B} + B\bar{C} + \bar{A}C$

3.10 $\bar{A}(\overline{BC})$

3.11 $\overline{ABC}$

3.12

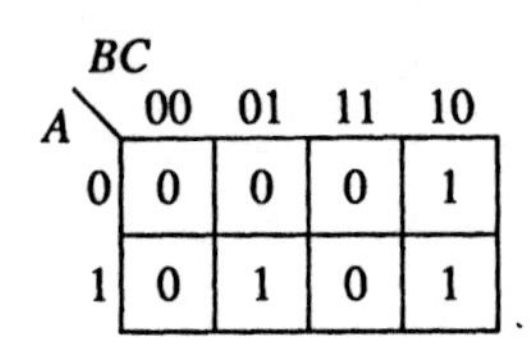

A \ BC	00	01	11	10
0	0	0	0	1
1	0	1	0	1

3.13 $A\bar{D}$

3.14 The function f_1 can be give as: $f_1 = \bar{A}\bar{C}D + \bar{A}C$. The group $\bar{A}\bar{C}D$ can then be shared in f_3.

3.15 By observation of the logic arrangement: $f = \overline{G_1 G_2A G_2B}$. Note the use of bubbles to show inversion on the inputs of gates.

3.16 $G_1 = 1, G_2A = 0, G_2B = 0$

Chapter 4

4.1

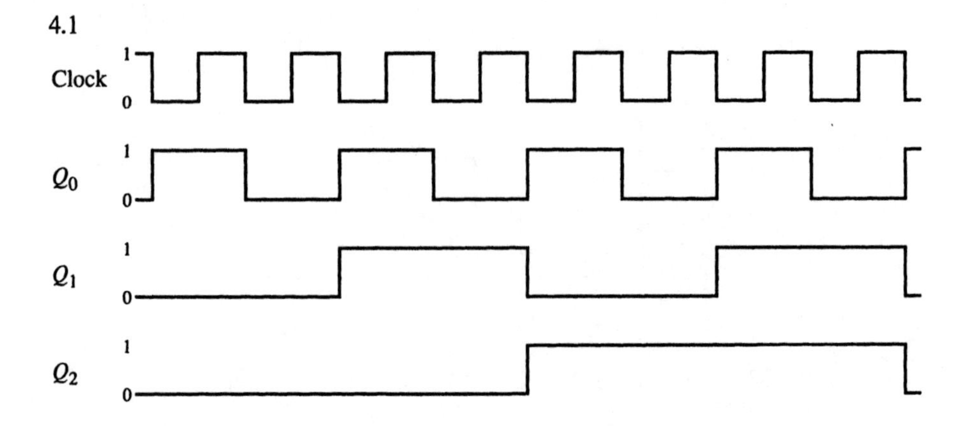

4.2 $Q = Q = 1$, i.e. a faulty situation.

4.3 From Figure 4.3, $Q = \overline{\bar{S}(\overline{\bar{R}Q})} = Q\bar{R} + R$ by applying DeMorgan's theorem. We can also say that $Q = 1$ if $S = 1$ (S sets output to a 1), or $Q = 1$ if already a 1 and $R = 0$ ($Q\bar{R}$). The characteristic equation can also be proved by forming a table listing the present values of S, R and Q and the new value of Q (Q_+). The function can be obtained from this table and minimized with a Karnaugh map. It is necessary to ensure that the new values of Q are consistent with existing entries with this value of Q as input.

4.4 From $Q_+ = J\bar{Q} + \bar{K}Q$ directly on the D-input, i.e.:

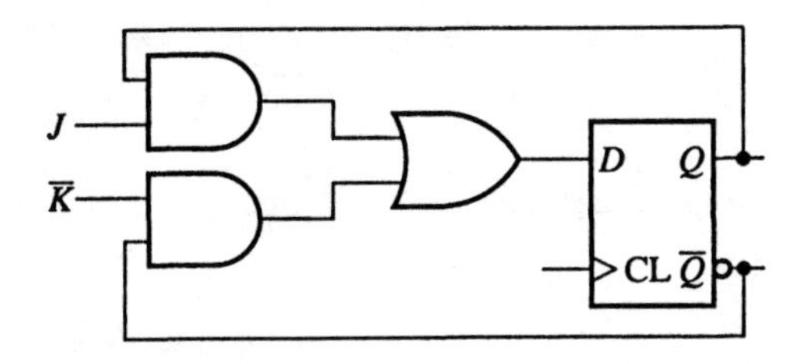

4.5 Given flip-flops $F_0, F_1, F_2, F_3, F_4, F_5, F_6$ and F_7, by connecting the reset inputs of flip-flops F_0, F_2, F_4 and F_6, and the set inputs of F_1, F_3, F_5 and F_7 together and to the (active-low) signal that causes the action.

4.6 The pattern 00000010 or 2 in decimal. Right shifts will round down the fractional part.

4.7 01010101 → 00101010 → 10010101 → 01001010 → 10100101 repeating.

4.8 Similar to a D-type implementation except the Q output of the final flip-flop connects to the K input of the first flip-flop and the $\bar{Q}$ output of the final flip-flop connects to the J input of the first flip-flop.

4.9 $J_A = K_A = 1$
$J_B = K_B = A$
$J_C = K_C = AB$
$J_D = K_D = ABC$
$J_E = K_E = (ABC)(D)$
$J_F = K_F = (ABC)(DE)$
$J_G = K_G = (ABC)(DEG)$
$J_H = K_H = (ABC)(DEG)H$

leading to the circuit overleaf.

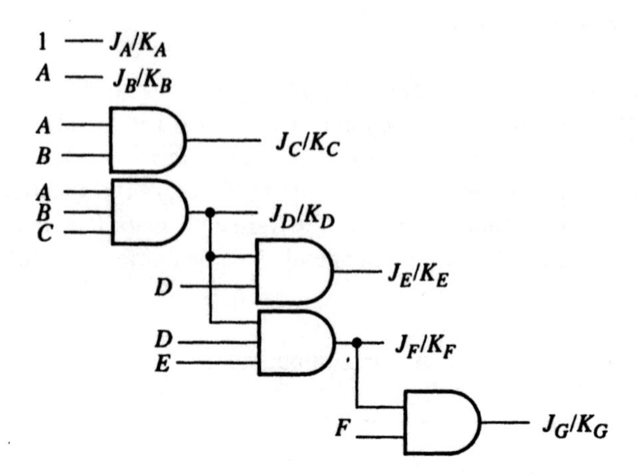

4.10 The Q output of one flip-flop to the clock input of the next flip-flop.

Chapter 5

5.1 Six (as $2^6 = 64$, the next higher power of 2).

5.2

Y_2 ($\bar{y}_2 y_1 \bar{C}$ circled at $C=0$, 01; $\bar{y}_2 \bar{y}_1 C$ circled at $C=1$, 00)

$C \backslash y_2y_1$	00	01	11	10
0	0	1	0	0
1	1	0	0	0

Y_1 ($\bar{y}_2 y_1 \bar{C}$, $\bar{y}_2 \bar{y}_1$, $y_2 y_1 C$ circled)

$C \backslash y_2y_1$	00	01	11	10
0	1	1	0	0
1	1	0	1	0

$$Y_2 = \bar{y}_2 y_1 \bar{C} + \bar{y}_2 \bar{y}_1 C$$
$$Y_1 = \bar{y}_2 y_1 \bar{C} + \bar{y}_2 \bar{y}_1 + y_2 y_1 C$$

5.3 $Y_1 = y_3 y_1 + y_3 y_2 + \bar{x}\bar{y}_2\bar{y}_1 + \bar{x} y_2 y_1$

5.4

1	00	00
2	01	10
3	11	11
4	10	01

5.5 While Z is a 0, it would depend upon the values of y_3, y_2 and y_1. Only if $y_3y_2y_1$

= 011 would the output function change and change immediately (from a 0 to a 1) (Look at the Karnaugh map of the output function.)

5.6 As usual, it is not possible to satisfy the rules completely. States *A*, *B* and *C* lead to a group of the same next states. States *A* and *C* also have the same output values. One possible assignment, using rules 1 and 3, would be to assign state *A* = 00 and state *C* = 01 (adjacent codes). States *A* and *D* in most cases also lead to the same next states and have the same output value, so we could assign state *D* = 10, leaving state *B* to have the assignment 11. Notice how reducing the number of states makes it more difficult to apply the "rules".

5.7

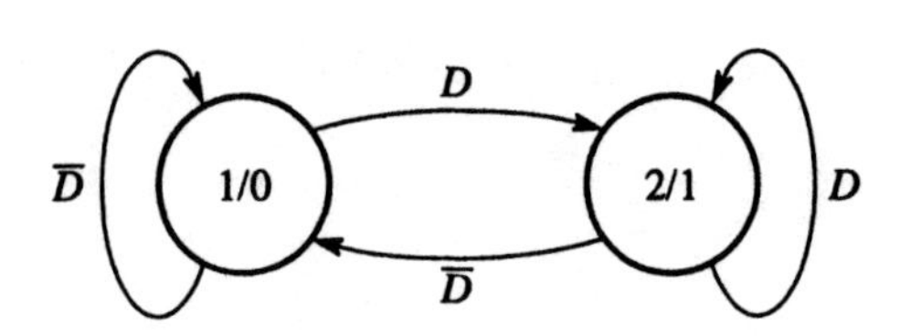

5.8

	y_0	y_1	y_2	y_3	y_4	y_5	y_6
State 0 =	0	0	0	0	0	0	0
State 1 =	1	0	0	0	0	0	0
State 2 =	0	1	0	0	0	0	0
State 3 =	0	0	1	0	0	0	0
State 4 =	0	0	0	1	0	0	0
State 5 =	0	0	0	0	1	0	0
State 6 =	0	0	0	0	0	1	0
State 7 =	0	0	0	0	0	0	1

Chapter 6

6.1

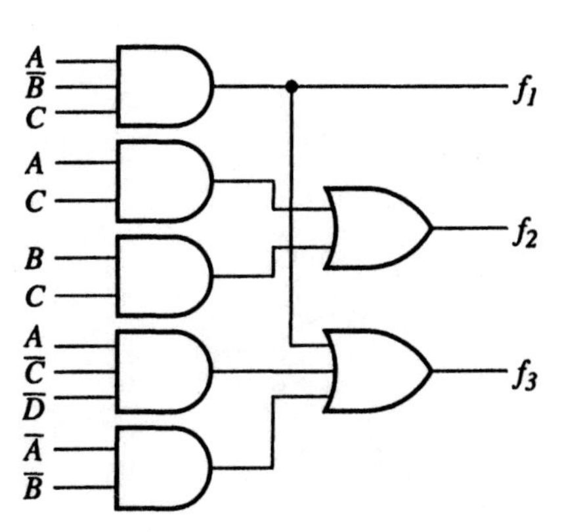

6.2 16 (10 inputs to the device plus 6 input/outputs that could be used as inputs).

6.3 Each output could have 7 product terms. There are 8 outputs. By using all the feedback connections, there could be $7 \times 8 = 56$ product terms.

6.4 $15 + 14 + 13 + 12 + 11 + 10 + 9 + 8 + 7 + 6 + 5 + 4 + 3 + 2 + 1 = (15 \times 14)/2 = 320$.

6.5 $Q_0 = \bar{A}\bar{B}C + A\bar{B}\bar{C}$
$Q_1 = \bar{A}BC + A\bar{B}\bar{C} + A\bar{B}C$
$Q_2 = \bar{A}\bar{B}C + \bar{A}BC + A\bar{B}C$

Given the don't cares, the minimized functions are:

$Q_0 = \bar{A}\bar{B}C + A\bar{C}$
$Q_1 = A + BC$
$Q_2 = C$

6.6 2^{17} (2 pins for power, 1 pin for the output leaving 17 pins for the address).

6.7 2^{10} (2 pins for power, 8 pins for the output leaving 10 pins for the address).

Chapter 7

7.1

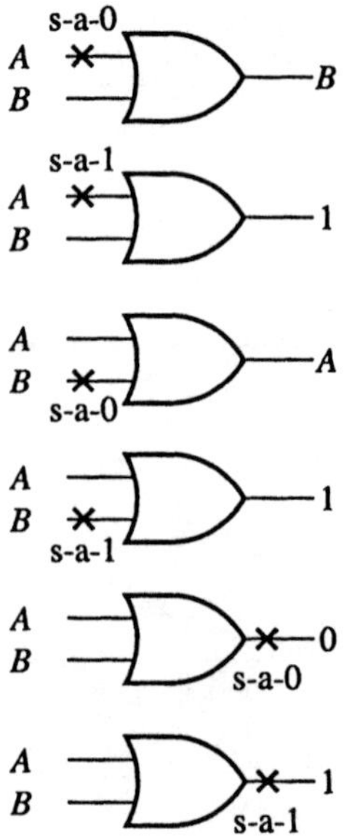

7.2 A logic 0 ($\overline{D}$).

7.3 NAND gate:
All s-a-0 input faults and the s-a-1 output are equivalent (test vector 111)
NOR gate:
All s-a-1 input faults and the s-a-0 output are equivalent (test vector 000)

7.4 NAND gate: 111 plus 011, 101, 110 (to test each input s-a-1).
NOR gate: 000 plus 100, 010, 001 (to test each input s-a-0).

7.5 Gate output were shared to reduce the number of gates in multi-output circuits.

7.6 011

Index